THE ROUGH GUIDE to

Genes & Cloning

Written by

Jess Buxton & Jon Turney

ROUGH
GUIDES

www.roughguides.com

Credits

The Rough Guide to Genes & Cloning

Editing: Matthew Milton
Picture research: Judi Burger
Typesetting: Andrew Clare & Michelle Bhatia
Proofreading: Martin Moore
Production: Aimee Hampson
& Katherine Owers

Rough Guides Reference

Series editor: Mark Ellingham
Editors: Peter Buckley,
Duncan Clark, Tracy Hopkins,
Sean Mahoney, Matthew Milton,
Joe Staines, Ruth Tidball
Director: Andrew Lockett

Cover picture credits

Front cover: Sheep portrait © Getty **Back cover:** Helix structure © Matthias Kulka/Corbis
Inside cover: ANDi the rhesus monkey © Mike Stewart/Corbis;
Fred Sanger portrait © Bettmann/Corbis

Publishing Information

This first edition published April, 2007 by
Rough Guides Ltd, 80 Strand, London WC2R 0RL
345 Hudson St, 4th Floor, New York 10014, USA
Email: mail@roughguides.com

Distributed by the Penguin Group

Penguin Books Ltd, 80 Strand, London WC2R 0RL
Penguin Putnam, Inc., 375 Hudson Street, NY 10014, USA
Penguin Group (Australia), 250 Camberwell Road, Camberwell, Victoria 3124, Australia
Penguin Books Canada Ltd, 90 Eglinton Avenue East, Toronto, Ontario, Canada M4P 2YE
Penguin Group (New Zealand), 67 Apollo Drive, Mairongi Bay, Auckland 1310, New Zealand

Printed in Italy by LegoPrint S.p.A

Typeset in DIN, Myriad and Minion

336 pages; includes index

A catalogue record for this book is available from the British Library

ISBN 13: 978-1-84353-759-5
ISBN 10: 1-84353-759-1

1 3 5 7 9 8 6 4 2

Contents

CONTENTS

Introduction

Scientists often need new words to help pin down new ideas. Mostly, this just adds to the jargon the rest of us don't care about. But a few of these words move out of the lab and take on a life of their own. That is a sure sign something interesting is going on, and "gene" is the most interesting example to have appeared in the last hundred years. Genes are the units of inheritance, the information which one generation passes to the next and which influence its characteristics. They are at the heart of life, of evolution and – more and more – at the heart of medicine, industry, and agriculture.

In the 20th century, researchers made astonishing leaps in understanding what genes are and how they work – the history of genes has been one of the greatest ever stories of discovery. But in the 21st, things are going to get increasingly personal. There will be more genetic tests available to offer clues to the diseases that you, or your children, might be vulnerable to. There will be new ways of treating or preventing these illnesses, tailored to your genetic make-up. Research on stem cells may well provide doctors with "**body repair kits**" – new cell-based treatments that could consign devastating conditions such as Parkinson's disease, or the paralysis that follows spinal injury, to the history books. There could be modified genes in the food you eat, while the same technology may one day be used to breed animals with organs suitable for transplanting into human patients. And as the century moves on, it is increasingly likely there will be widespread efforts to "improve" species – and inevitably the human race – by enhancing genes. Whether or not this would be a welcome development, it would truly be a new phase in the evolution of life.

However, to recall one celebrated 20th century slogan, the personal is also political. Who is going to pay for all this stuff? Who will regulate these new technologies? Who will decide whether anyone should in fact be doing any of these things at all – whether it's commercial biotechnology companies wishing to patent the genes they have mapped, or research that uses human embryos to discover how stem cells work? And should we worry that the choices that individuals make for themselves and their offspring – however well intentioned – might end up ushering in a brave new world of **eugenics**?

Getting a handle on these quandaries is both helped and hindered by the vast amount of information about genes that now permeates our culture. Some popular accounts claim that genes are all-powerful, the blueprint that shapes us, and the root of all the instincts that determine our behaviour. Then there are the science fiction comic strips and movies that depict superhumans, soulless clones and monsters. The fanciful powers of mutant heroes **The X-Men**, or Bruce Banner's shirt-ripping transformation into **The Incredible Hulk**, are all something do with genes – so the scriptwriters suggest. Meanwhile, advertisements tell us that there is DNA technology in hair shampoo and that we can buy a thoroughbred car with superior genes. And there is the constant flow of scientific reports, and the news stories which follow them, promising that genetic discoveries will bring new ways of tackling everything from obesity and cancer to schizophrenia and depression, and will also feed the world, eradicate pollution and boost our economies all at the same time.

This guidebook, like any guidebook, is an invitation to further exploration. We want to arm you for that exploration with a good sense of what the scientific world already knows about genes, what is possible with genetic and cloning technology, and whose claims about what might happen next make sense.

There's a lot of fascinating science involved. But this is a book which tries hard to avoid the pitfalls of many popular science tomes. One is to get so caught up in the science that it appears in a cultural vacuum, emerging from the laboratories without any clue as to its context in the rest of the world. All too often, ninety percent of a text will provide a neat exposition of the marvels of science, only dealing with any social issues that the science throws up in a hasty afterword. Well-intentioned scientists often do this, and conclude with some sonorous statement that the uses of science are "up to all of us to decide". That's true. But we might need a bit more help than that. So when ethical or political issues arise– and in this science that happens a lot – we try to flag them up clearly. And we look critically at the larger narratives people have constructed in order to come to terms with the implications of genes, whether they are presented as fact or fiction. We don't think, like some postmodernists, that science is just a story. But scientists nevertheless do tell stories. What makes science so powerful is that they sometimes manage to make them come true.

Acknowledgements

Jess Buxton would like to thank her colleagues at Progress Educational Trust – especially Kirsty Horsey, for her many helpful comments, and Marcus Pembrey for his continued support and infectious enthusiasm. Special thanks to Pat and Mary for all their help and understanding, and also to Liam, Lucy, Robert and Maddy. Finally, thanks to Jon for the opportunity to work with him on such an interesting project, and for all his invaluable advice along the way.

Jon Turney thanks colleagues and students in the Department of Science and Technology Studies at University College London, where he taught about – and talked about – genes a lot in the 1990s. Thanks, too, to all the editors who have allowed him to write about genes then and since. Closer to home, thanks to Danielle, Catherine and especially Eleanor, who gave up part of her summer for research assistance. And thanks to Jess for being a brilliant, expert and efficient collaborator.

About the authors

Jess Buxton is a geneticist and science writer based at Imperial College, London. She is also genetics editor of *BioNews* (www.bionews.org.uk), which provides news and comment on human genetics and assisted reproduction technologies.

Jon Turney is a science writer and lecturer in London. He has degrees in biochemistry and history of science, and his books include *Frankenstein's Footsteps: Science, Genetics And Popular Culture* (1998). He is course leader for the MSc in Creative Non-fiction at Imperial College London. He is devoted to understanding the latest developments in molecular biology and to watching monster movies.

PART 1
THE BASICS

FAQs

Key questions and answers

Before exploring the various aspects of genes and cloning in depth, let's quickly answer some of the most frequently asked questions about the issues. The following pages will bring you up to speed with the current situation and some of the future possibilities, before the rest of the book delves deeper.

The small picture

What is a gene?

Genes are the basic biological unit of inheritance, the information that is passed on from one generation to the next. We have two sets of around 25,000 different genes (one set inherited from each parent), which together influence our appearance, health and maybe even aspects of our personality. Genes are made out of DNA, and most are coded instructions for making everything your body needs to develop, grow and survive.

And what exactly is cloning?

Cloning has two distinct meanings. The first describes the creation of an exact genetic replica of another living organism (for example Dolly the sheep) or a single cell. Growing new plants from cuttings is a form of cloning too, and identical twins are naturally occurring clones. The second definition of cloning refers to the copying of a single piece of DNA (which may or may not be a gene), by propagating it in a bacteria or yeast.

Has anyone cloned a human yet?

No – though several dubious claims have been made to the contrary. In fact, scientists have not yet managed to clone any primates, leading some to suspect that it might not even be possible to do so using current methods. Whatever the reason for the technical difficulties, the problems and low success rate associated with cloning have led to human cloning being outlawed in many countries. Some, including the UK, do allow scientists to create cloned human embryos – so-called "**therapeutic cloning**" research – so long as they are destroyed before they are 14 days old, and are not implanted.

Can I clone my pet cat?

Not very easily. You would need some well-preserved tissue from your cat, for a start – say, a frozen piece of skin. Then you would need to find a scientist or company offering the service. US pet cloning firm **Genetic Savings and Clone** recently stopped taking new orders, because it was "unable to develop the technology to the point that cloning pets was commercially viable". Although it produced the first commercially cloned kitten, "Little Nicky", born in October 2004, the company cloned a total of just five cats. And, like most other cloning scientists, they never did get dog cloning to work. (See p.237 for information about **Snuppy**, the world's first cloned dog).

Can genetic testing tell me what illnesses I'll get when I'm older?

For most common illnesses, such as cancer, heart disease and Alzheimer's disease, the answer is usually no. This is because these diseases are influenced by a whole host of different factors – diet, lifestyle, exposure to toxins - not just genes. However, genetic tests are available for families affected by genetic conditions that develop in adulthood, such as Huntington's disease, and some rare forms of inherited cancer.

How much to get my genetic code read?

The first draft of the entire human genetic code, or **genome**, published in 2003, cost an estimated $3 billion. Using the same methods, it would set you back a staggering $10–$50 million to have your own genetic code read at present. But the US **National Institutes of Health** (NIH) and others are funding research into new, cheaper technologies, aimed at developing a "$1000 genome" by 2014. Many of those in the race think that the $100,000 genome will be with us in the next few years.

Does my DNA reveal my ancestry?

DNA testing can often provide clues about ancestry, but not definitive answers. It can help determine if two people are related, for example, or if they share a common ancestor. Two types of test exist: those that look at **mitochondrial DNA**, which is passed on from a mother to all her children, and Y-chromosome DNA, which is passed from father to son. So mitochondrial DNA tests can provide clues about ancestors on your mother's side of the family, while men can also track their male ancestors.

Does my doctor care about my family's medical history?

Definitely – your family history can provide your doctor with important information about your future health. Illnesses such as heart disease, schizophrenia and diabetes are all influenced by genes, and so often 'run in families'. Information about which of your family members is affected can help doctors estimate your own risk, although there are no genetic tests available for these illnesses (except for some rare inherited forms).

Should I take part in a medical database project?

There's no reason not to take part in projects such as the UK's Biobank (who, in their own words are a "unique resource for ethical research into genetic and environmental factors that impact on human health and disease") or similar efforts worldwide that are attempting to understand the links between genes, disease and health – although you'd be improving health care for future generations, rather than for yourself. These large-scale projects probably won't bear fruit for at least another ten to twenty years, but the results should eventually lead to earlier diagnosis and better treatments for many common illnesses.

Do I have to tell my insurance company my genetic test results?

It depends on where you live. In the UK, a moratorium is currently in place to restrict the use of genetic test results by insurers (which is due to expire in November 2011).

The agreement states that no one will be required to reveal the results of a genetic test unless it is first approved by the **Genetics and Insurance Committee** (GAIC), and unless it is for insurance of more than £500,000 (for life insurance), or £300,000 (for critical illness and income protection insurance). So far, GAIC have only approved the use of one genetic test - that for **Huntington's disease** (HD), an incurable brain disorder that first strikes during middle age.

Do the police have my DNA "profile"?

Again, it depends on where you live, and whether you've ever been held in a police station. In the UK, police can take non-intimate samples – usually a cheek swab – without consent from anyone who finds themselves in a police station during investigation of a "recordable offence", basically anything more serious than a traffic violation.

The sample can be taken even if it is not relevant to the crime under investigation. Currently, the UK police's **National DNA Database** (NDNAD) holds the DNA profiles of over three million people – more than any other country in Europe.

Can I get a paternity test online?

Yes, there are now many companies offering cheap DNA tests to establish paternity – most of which simply require a cheek swab, blood sample or hair follicle from the child and father. However, concerns continue to be voiced that many paternity tests are offered with no counselling or follow-up on what, for many, could be disturbing news. Exact figures are hard to come by, but there may be 20,000 or more paternity tests annually in the UK, and 100,000 in the US.

Is there a "gay gene"?

No, although studies of the families of gay men show they are more statistically likely to have gay brothers, uncles and cousins than are heterosexual men. Studies of twins also point to a possible genetic influence on sexuality, although genes cannot of course be the whole story – even in the case of identical twins, the chances that one twin will be gay if his brother is are around 50-50.

Despite the fact that no "gay gene" exists, some have claimed that there is evidence to support an underlying biological basis for homosexuality, though it's unclear whether this is the effect of genes, other factors that affect the developing foetus, or an interaction between the two.

Could I have a "designer baby"?

Doctors can test embryos created using *in vitro* fertilisation (IVF) for the presence of gene mutations that cause disease, in order to select (not design or alter) unaffected embryos to return to the womb. This technique, called **pre-implantation genetic diagnosis** (PGD) is expensive, invasive and has a low success rate, so it is only really used by couples who have a high risk of having children who might be affected by a serious genetic condition.

It will probably never be possible to select "designer babies" with a shopping list of characteristics such as hair colour and sporting ability - even if there were ever a demand for such children.

Will my children be able to genetically "enhance" their children?

Unlikely, but it may be possible for their grandchildren, or great-grandchildren to do so: the real question is, *should* they? The techniques for altering the genetic make-up of plants and animals are now fairly routine, but creating "**GM humans**" by altering genes in eggs, sperm or embryos (so-called **germ-line gene therapy**) is outlawed by every country that has legislation in this area. At the moment, scientists just don't know enough about what different genes do in the body, or how they interact with each other. So trying to alter a single gene could have multiple, unpredictable and potentially disastrous effects.

Could GM food help the developing world?

GM technology can be used to alter the genetic make-up of crops in many beneficial ways, some of which are already in use. Research into creating crops with improved resistance to viruses and fungal attack is ongoing, as is work aimed at making plants more resistant to drought.

Scientists are also trying to improve the nutrient content of crop species such as rice, wheat, maize, cassava, millet, and potato – all advances that could potentially help people living in the developing world. But introducing new crops into any country first requires a careful assessment of its economy and agricultural system, and the available alternatives. There are, of course, many contentious political and economic issues that take the GM-crops debate well beyond the science of what they can do.

Could *Jurassic Park* really happen?

In the book and subsequent film *Jurassic Park*, whole dinosaurs are recreated using DNA from dinosaur blood preserved in the guts of prehistoric insects trapped in amber. In reality, even DNA – an unusually stable molecule – cannot survive for the millions of years necessary to make such a thing a reality. Short fragments of DNA may last for 50,000 to 100,000 years before water, air and microbes finally destroy them. So, sadly for dinosaur fans, *Jurassic Park* is, and will in all likelihood remain, just fiction.

Cells and proteins

What is a cell?

Getting to grips with genes and cloning requires a close look at the stuff of life. All living things are made up of cells (apart from viruses, that is, which only "live" by infecting cells). Some life forms, such as amoebae, are made of just one cell. There is a tiny soil-dwelling worm called *Caenorhabditis elegans* that is made of exactly 959 cells, while humans are made up of about 100 million million cells.

Like all other multicellular creatures, humans begin life as a single cell: a fertilized egg smaller than a full stop. This divides into two, then four, then eight – and so on. Your body has around two hundred different types of cell in it, all working together day and night. Each cell must make the substances it needs to survive, grow, multiply and do its job. Every cell is

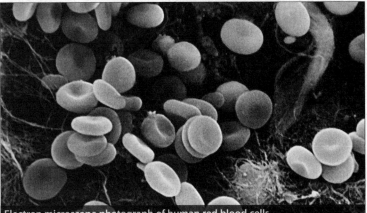
Electron microscope photograph of human red blood cells

The cell theory of life

The first person to use the word "cell" was the English scientist Robert Hooke. In 1665, he was looking at thin slices of cork under his microscope and saw compartments that reminded him of the cells found in monasteries and prisons. Hooke's detailed drawings of these and many other microscopic observations appeared in his most famous book, *Micrographia*. Ten years later, Dutch scientist Antonie van Leeuwenhoek discovered microscopic single-celled creatures living in water. In 1678, when van Leeuwenhoek wrote to the Royal Society (the UK's long-standing scientific academy) reporting his discovery of these "little animals", it was Hooke who confirmed the findings. However, it was nearly two hundred years after these first observations that cells were recognized as the fundamental building blocks of all animal and plant life. This realization followed the advent of more powerful light microscopes, and staining techniques that allowed better visualization of the tissues viewed under them. Eventually, between 1838 and 1839, Germans Matthias Schleiden and Theodor Schwann argued that all living creatures are made up of cells, and that a cell is the smallest unit of life: the cell theory. The theory gained widespread acceptance some twenty years later when another German, called Rudolf Virchow, completed the picture – proposing that all cells arise from the division (splitting) of an existing cell, or *omnis cellulae e cellula* ("All cells arise from cells").

German scientist Rudolf Virchow, c.1900

a specialized factory, producing exactly what it needs to maintain itself. It must also communicate and cooperate with other cells to form tissues and organs, and to do everything necessary to keep the body alive and healthy – red blood cells, for example, need to carry oxygen to every part of your body. Cells will even sacrifice themselves, if required. Understanding how cells interact and work forms the basis of much modern biology – and indeed, of medicine. This book will focus on just one part of a cell's daily business, but it is an absolutely crucial one: how it stores and processes biological information, and how scientists have used this knowledge to manipulate life itself.

Different types of cell

There are two main types of cell: prokaryotic and eukaryotic. Prokaryotic cells have a single, circular strand of genetic material that floats around freely, unlike the genetic material of a eukaryotic cell, which is housed in its nucleus. All bacteria are prokaryotes, as are archaea – creatures that look like bacteria, but which have very different genes. Many archaea live in extreme environments where nothing else can survive, such as in hot springs or very salty water. Amoebae, yeast, plants, humans and other animals are all eukaryotic life forms. There are other differences between the two cell types: prokaryotes are generally much smaller than eukaryotic cells, and lack their highly organized internal structure. It is this internal organization, and resulting complexity of eukaryotic cells that has allowed them to specialize, and form the many different cell types that make up the tissues of higher lifeforms. So-called "multicellular organisms" are all made up of eukaryotic cells, while prokaryotes are single-celled organisms, though they can form aggregates known as colonies.

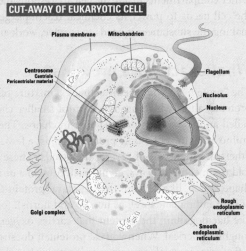

CUT-AWAY OF EUKARYOTIC CELL

Plasma membrane Mitochondrion
Centrosome
Centriole
Pericentriolar material
Flagellum
Nucleolus
Nucleus
Golgi complex
Rough endoplasmic reticulum
Smooth endoplasmic reticulum

Inside a cell

Some cells are large enough to be visible to the naked eye – an unfertilized ostrich egg, for instance, is a single cell that weighs some 3.3 pounds. But to see most other types of cell, you need a microscope. Under low magnification, most cells look like simple blobs, but closer examination reveals a highly organized, compartmentalized structure. Every day, thousands of different chemical reactions take place in a cell: without this order biological anarchy would ensue. The raw materials for all these chemical reactions come from the food you eat: cells are a mixture of water (which makes up about two thirds of every cell), proteins, carbohydrates, fats and tiny amounts of minerals. The body is essentially a vast, complex recycling plant, churning up and breaking down everything you put in, and turning most of it into something useful.

Whilst all of the components of a cell are essential, it's the proteins that are arguably the most interesting, since they are the key to life's diversity – different types of cell use, and produce, different proteins. At the centre of almost every cell in your body is the nucleus. If cells are like factories, then the nucleus is its "archive store" – where the designs for everything made in the factory are kept. The nucleus actually contains a copy of your genes, which are instructions for making the all-important proteins. Other compartments inside the cell are responsible for making the energy the cell needs to power its chemical reactions, getting rid of waste, and making the substances it needs to survive, work and grow.

What exactly are proteins?

The "protein" in your diet is actually a mixture of thousands of different proteins, all of which are made up of smaller chemical building blocks called amino acids, joined together in chains. There are twenty different amino acids in cells. Some proteins are just a few amino acids long, while others are made up of several thousands. These chains of amino acids fold up in complex ways, giving each protein a unique three-dimensional shape. Some, like keratin, the structural protein that makes up the bulk of hair and nails (as well as hooves and horns), form long fibres. Others, like haemoglobin, the protein that carries oxygen around the body, are roughly spherical. When you eat protein, it doesn't matter where it's from – fish, cheese, steak or soya – it all gets broken down into its constituent amino acids, which your cells then use to make their own, custom-built proteins. This is good news for vegetarians, since it means that they don't need to eat meat to make, say, muscle proteins. (Although they do need to make sure they eat a variety of different non-meat proteins to get enough of all nine "essential" amino acids, which the body can then use to make the other eleven "non-essential" amino acids).

How do cells make proteins?

We have known roughly how cells make proteins since the 1960s, which saw a flurry of activity following the discovery of the DNA (deoxyribonucleic acid) double helix. Molecular biology pioneers such as James Watson, Francis Crick, Sydney Brenner and others worked out that genes are pieces of DNA, and crucially, that DNA contains the coded information used by cells to make proteins. This is not to say there's nothing left to find out – far from it. Scientists think that humans have at least 300,000–400,000 different proteins, but what many of them actually do within the body remains a mystery. There's also the question of why we seem to have far more proteins than genes – humans have a "mere" 25,000 genes (at the last count). What follows is an overview of what we *do* know, based on the results of countless experiments and years of painstaking observation.

DNA is unusually stable for a biological molecule, and under the right conditions can survive intact outside the body for years, occasionally millennia – a property invaluable to scientists studying scraps of genes from ancient humans. It is this stability, along with its ability to act as a template for making copies of itself, that makes DNA the ideal molecule for storing and passing on information, generation after generation. If you could trace your DNA back through all your ancestors – human and otherwise – you would find that bits of it were shared by some of the very first creatures to live on earth. The unique properties of DNA are down to its structure, the famous double helix.

Francis Crick (1916–2004)

Francis Crick was the elder of the dynamic duo of the double helix, his scientific career interrupted by a World War II stint designing mines for the British Admiralty. He trained as a physicist, but switched to biology after the war. Crucially, he had a profound knowledge of the maths needed to interpret X-ray photographs of complicated crystal structures.

After his great success with Watson in cracking the structure of DNA, he went on to be the leading theorist of the new molecular biology. His work on the genetic code and protein synthesis was central to the new science, but he also worked on virus structures and on the origin of life. He shunned public honours and administration, and worked for the British Medical Research Council in Cambridge until regular retirement age, when he moved to the Salk Institute in California and switched to neuroscience for the last twenty years of his life.

Although once described by the acerbic biochemist Erwin Chargaff as looking like "a faded racing tout", Crick had an outgoing personality and a fertile, critical mind, which were the basis for several successful scientific collaborations: with Watson, with Sydney Brenner, chemist Leslie Orgel and, latterly, theoretical neuroscientist Christoff Koch. He always stressed that being a successful theorist meant more than having new ideas. It needed a willingness to discard them and have new ones, over and over. He told his own version of the double helix story in his autobiographical *What Mad Pursuit?* and outlined his ideas on consciousness in *The Astonishing Hypothesis*.

James Watson (1928–)

After a childhood career as a precocious TV quiz kid, Watson went to the University of Chicago at 15, and joined Crick in Cambridge in late 1951. Following the pair's success with the double helix, he worked on viruses and RNA, and helped develop strategies for cancer research. Unlike Crick, he spent more and more time on administration and policy, first at Harvard, then as chief of the famous Cold Spring Harbor Laboratory in New York and as the first director of the Human Genome Project.

Famously outspoken – Darwinian and naturalist E. O. Wilson described his young Harvard colleague as "the most unpleasant human being [he] had ever met" – he reinvented the scientific memoir with *The Double Helix* in 1968. It was just one of a string of books which added up to what basically constituted a second career as a writer. They include the bestselling textbook *Molecular Biology of the Gene*, a market leader for forty years, and a belated sequel to *The Double Helix*, titled *Genes, Girls And Gamow*. It was less compelling than the earlier book, and reinforces the feeling that, somehow, nothing else in Watson's long and distinguished career has quite lived up to those great days in Cambridge, England.

Discovering DNA: the double helix

"If the apple doesn't fall, let's go shake the tree." The screenwriter William Nicholson put these words into the mouth of James Watson (as played by Jeff Goldblum: intense, rude, American) as he talked about how to tackle the structure of DNA with fellow biologist Francis Crick (as played by Tim Piggott-Smith: loud, fast-talking, English) in the 1987 BBC film drama *Life Story*. Their great discovery of 1953 has become one of the icons of twentieth-century science, and the story of how they did it is almost equally celebrated.

The film, which tells a version of the DNA story close to Watson's novelistic non-fiction classic *The Double Helix*, depicts one stereotypical perception of scientists and how they work: obsessing about important problems and worrying about the competition in a fraught process full of doubts and fears, punctuated by moments of sublime inspiration. Was their discovery really like that? Probably not – Francis Crick commented that Watson's book obscures how much hard work they did. He himself never once thought there might be a Nobel prize in the offing. And although Watson portrays the pair as preoccupied by a race to find the answer with the American chemist, **Linus Pauling**, Crick did not see it that way.

But by the early 1950s Watson and Crick did share a crucial conviction that DNA was important. That importance only emerged extremely gradually in the years after the substance was first identified back in 1869 by **Friedrich Miescher**. In extracts of cell nuclei taken from pus-soaked bandages, Miescher had found a phosphorus-rich substance that he called nuclein. He showed that it had a basic protein portion, now recognized as part of the packaging of chromosomes, and an acidic portion – DNA.

Friedrich Miescher

When the idea of genes came along thirty years later, the chromosomes in the nuclei of cells seemed to contain the genes – but what were they made of? Initially, attention focused mainly on the *proteins* as the candidates for the genetic material. DNA seemed to be a rather boring molecule, not complex enough to do more than hold a chromosome together. Proteins were wildly various, so surely they must be the key to understanding heredity?

That did not finally change until as late as 1944, when an American medical researcher called **Oswald Avery** completed experiments on bacterial gene transfer that proved beyond doubt that DNA carries genetic information. Some kinds of bacteria conjugate, in effect having bacterial sex. A tube connects one cell to another, and material passes between them. Avery showed that the recipient's genes can be altered if – and only if – DNA is transferred.

This meant that as science began to pick up speed after World War II attention began to turn to the structure of the DNA molecule. Ideas from earlier work on proteins fed in to the new problem. **Linus Pauling**, working in California at CalTech, had produced startling insights into some of the most common features of protein structures. His ideas were fuelled by results derived from the imaging technique known as X-ray crystallography, and he developed them by building models that scaled up the sizes and distances between atoms to things you could heft in your hand. One of his most inspired moments came when, bedridden with flu, he roughly

Maurice Wilkins (1916–2004)

Maurice Wilkins was born in New Zealand and came to England as a boy. Like Francis Crick, he trained as a physicist and worked for the war effort, in his case on the Manhattan project to make the atomic bomb. And like Crick he switched to biology. But he was unlike Crick in other ways – shy, self-effacing and uneasy with women. His crucial role in solving the DNA structure came from his expertise in X-ray crystallography, and his early images of DNA caught the attention of James Watson.

He continued working on DNA at King's College, London after the double helix breakthrough – his colleague Rosalind Franklin had already left by the time the classic paper was published. Wilkins' later work verified many details of the structure. He was also a committed peace activist both before and after the war, and he lent his name to the fledgling British Society for Social Responsibility in Science in the 1960s, remaining its president for many years. He continued teaching at King's after his official retirement, again focusing on science and society. After much hesitation, he wrote his own account of his life and his contribution to the DNA story in *The Third Man Of The Double Helix*. The book was finally published in 2003, fifty years after the discovery, and just a few months before his death.

drew atoms, bonds and angles on a piece of paper, folded the paper, and found that one regular sequence of amino acids could be twisted into a helix – a long drawn out spiral structure. Could DNA be a helix, too?

Two ambitious young men in England, Francis Crick and the visiting American James Watson, met in the Cavendish Laboratory in Cambridge in 1951 and quickly agreed to work together on DNA. Crick knew about X-ray diffraction, Watson about genetics and viruses. Together they reckoned they might unlock the structure of what they both fervently believed was the most important molecule in biology.

Like Pauling, Watson and Crick decided to build models, using what was then known about the molecule's chemistry and composition, as well as X-ray images from DNA crystals or fibres. The idea behind X-ray crystallography is that anything that can form crystals must have a regular structure, no matter how complex it is. A beam of X-rays passing through a crystal should make a pattern on a photographic plate which is related to the position of atoms in the molecule. But how the two were related had to be worked out by intricate mathematical analysis. Hands-on model building might offer a short cut.

Their early efforts faltered, from a combination of knowing too little chemistry and reliance on poor X-ray data. One early effort, a three-chain

Rosalind Franklin (1920–58)

Rosalind Franklin was building a reputation as a resourceful investigator of chemical structures, with fruitful research stints in Cambridge and Paris, when she joined King's College, London in the same research group as Maurice Wilkins. There she took up work on DNA, with enduringly controversial results. Mixed messages from the lab's director led both her and Wilkins to think DNA was their personal project, though Wilkins was already working on the problem, and communication – let alone collaboration – between them was poor to non-existent.

She took the best X-ray photos of DNA then seen by anyone, and it was Watson's glimpse of these, aided by Wilkins, which helped him and Crick deduce crucial aspects of the double helix structure. Debate has gone on ever since about whether she was close to solving the problem herself, though it is not clear that she was as attached to the biological importance of DNA as her colleagues.

Soon after the DNA episode, she left the rather old-fashioned atmosphere of King's for another London lab, at Birkbeck College, where she did important work on the structure of viruses. She had a tense relationship with Watson, and was caricatured for dramatic effect in Watson's famous memoir. However, she became good friends with Francis Crick, who respected her skills in the lab. She died a few years later of ovarian cancer, aged just 37, so did not live to see the Nobel prize award to Watson, Crick and Wilkins. There are two biographies, however. Anne Sayre's *Rosalind Franklin And DNA* claimed her as a feminist heroine. Brenda Maddox's more recent *Rosalind Franklin: The Dark Lady Of DNA*, is more measured, and a better read.

helix with the chemical "backbone" of sugars and phosphate groups on the inside of the molecule, was quickly shot down when examined by visitors from King's College London – **Maurice Wilkins** and **Rosalind Franklin** – who were also doing X-ray studies of DNA.

Franklin was working by herself at the time, having been given responsibility for an ongoing DNA project at King's. She and Wilkins never really got on after their lab boss created confusion about which of them was working on DNA. But she was highly skilled in crystallographic analysis, and was taking the best X-ray photos. In 1952, she found that images of the "wet" form of DNA suggested a helical structure, something she had previously doubted. Wilkins shared these photos with Watson, and he and Crick saw that the symmetry of the image indicated a double helix. They returned to model building, which Franklin thought unhelpful.

That, at least, is the simple version of the oft-told story – a quick look at someone else's photos triggering the Cambridge pair's breakthrough. But Watson and Crick took a further crucial step when they read a report from the King's College lab to the British Medical Research Council, which detailed measurements from Franklin's observations. Ironically, the report was designed to encourage collaboration between the London and Cambridge labs. It told the Cambridge duo that there were two strands of DNA wound round each other and, for the first time, that they ran in opposite directions. Watson built on this by playing around with carefully made scale models of the four chemical units – the "bases" **adenine**, **thymine**, **guanine** and **cytosine**, known for short as A, T, G and C – which are linked together in the two chains of the DNA structure. Juggling them around, he found that if the bases had to be inside the helix, then they formed two matched pairs. Their chemical structures meant that A fit snugly with T and G with C.

This beautiful finding immediately suggested how the DNA strand might store information, and how it could be passed on. If you unwound

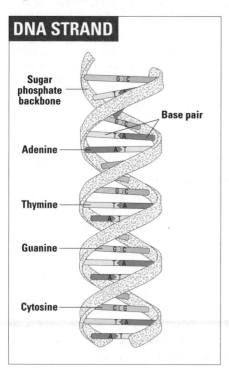

and separated the strands, each could act as a template for making a new copy of the other. There was no room to spell this out in Watson and Crick's rapidly drafted initial letter to the journal *Nature* in April 1953, but they did take care to say in almost the last sentence that: "it has not escaped our notice that the specific pairing we have postulated imme- diately suggests a possible copying mechanism for the genetic material." (Should you wish, you can read the whole paper online at www. nature.com/nature/dna50 /archive.html)

It's hard to believe now, but DNA had a quiet debut. A great deal of research fol-

DNA STRAND

Sugar phosphate backbone

Base pair

Adenine

Thymine

Guanine

Cytosine

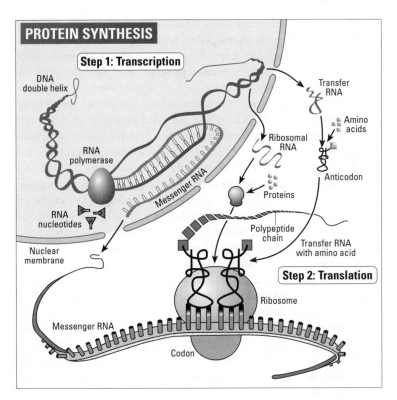

lowed to work out how the information in DNA gets used. The idea that this paper – for which the order of the authors' names was decided on a coin toss – was the foundation stone of a whole new science only took hold gradually. But by the end of the twentieth century, after breakthroughs such as recombinant DNA and DNA fingerprinting, and the Human Genome Project, DNA was the most famous molecule in science. And the double helix icon could be found everywhere from share prospectuses to cosmetics adverts. Not bad for stuff which began its career as pus.

Using DNA

The discovery of the structure of DNA was a great achievement, but only the start of a huge amount of research into how the stuff is actually used by living systems. How exactly do cells use the coded DNA information to make proteins?

Each cell makes only the proteins it needs – hair follicle cells make lots of keratin but don't make haemoglobin, for example. So although nearly all your cells have a complete copy of the entire set of human gene instructions (collectively termed the **human genome**), each cell uses only the instructions it needs. For a cell to make a particular protein, it first needs to access the relevant gene from the estimated 25,000 different human genes contained in its nucleus. These genes are packaged up in 46 bundles called **chromosomes** (derived from the Greek for "coloured bodies", because they were first identified under the microscope as structures that absorbed a lot of the dye used to stain cells). Chromosomes are like filing cabinets, while genes represent individual designs. Each chromosome is actually a single length of DNA, very tightly packaged and coiled up around proteins. An oft-quoted statistic is that if all of the DNA in just one human cell were to be stretched out, it would be almost two metres long. Instead, it is all bundled into the cell nucleus, which has a diameter of just five thousandths of a millimetre. Highly efficient packaging.

Reading a gene instruction entails unraveling the part of the chromosome where that gene is located to expose the required piece of DNA code. The cell then makes a "copy" of the instruction, using a close chemical relative of DNA called **RNA** (ribonucleic acid) – just as someone might photocopy an important file rather than remove the original from where it is stored. The RNA copy is then whisked out of the cell nucleus into the cell's **cytoplasm** (the jelly-like substance that fills cells), where the coded information is "translated" into proteins.

To sum up, then, genes are made of DNA, which is packaged up as chromosomes in the cell nucleus. Genes are the instructions for making proteins, which the cell "reads" using an intermediary, RNA. This "DNA-RNA-protein" flow of information was described by **Francis Crick** as the "central dogma" of molecular biology. He meant that the information can go only one way – any changes to proteins that occur within the cell's lifetime cannot alter its stored DNA code. So the message that gets passed on to an organism's offspring usually remains stable over many generations. Although there are some recently revealed exceptions to this, which we discuss later – in higher organisms the details get very complex – this basic idea that genes are used to make RNA, and that RNA is used to make protein is still a fundamental fact about the cell. The next section describes this whole process in a bit more detail.

Cracking the code

Although it is described as a double helix, it's perhaps more useful to think of a DNA molecule as a "twisted zip". Like a zip, it can be undone, revealing the "teeth" that make up each half. The "teeth" of DNA are actually individual "letters" of the DNA code which, once exposed, can be used by the cell to make the RNA copy. This happens every time a cell needs to make a particular protein, so many parts of the DNA code are usually "unzipped" at any one time – a stretch for every protein being produced. DNA also unzips when it needs to make a near perfect DNA copy of itself, as happens when a cell divides to produce two new cells.

Bringing DNA out of the dark: Francis Crick (pictured here lecturing at Cambridge, 1979)

So what is the exact nature of the DNA code? It turns out to be surprisingly simple, given the vast and complex array of proteins that a cell needs to make. Strands of DNA are made up of four different chemical building blocks, or **bases** – adenine, cytosine, thymine and guanine. These are the "letters" of the DNA code, A, C, T and G. These bases are attached to a chemical backbone, forming a strand of DNA, and two strands wind around each other to make the twisted zip. So each set of interlocking "teeth" is actually a base-pair of DNA: every A is paired with a T, and every C with a G. DNA molecules are measured in base-pairs – an average human gene can be anything from a few hundred to a few thousand base-pairs in length. The units **kilobases** and **megabases** are more often used to describe big chunks of human DNA, as the scientists who pieced together the map of the entire human genome spent many hours doing.

RNA is similar to DNA: it consists of strands of four different chemical building blocks, or bases, attached to a chemical backbone. But unlike DNA, in which two strands coil around each other, RNA is a single strand. Like DNA, RNA has the bases A, C and G, but instead of thymine (T), it uses another base called **uracil** (U) – which, like thymine, can pair up with the complementary base adenine. All proteins are made up of combinations of twenty different amino acids, but the code for both RNA and DNA is written in just four different chemical "letters". How can a code with just four letters

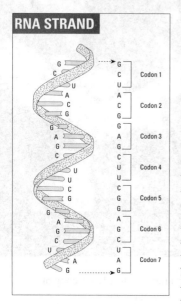

be used to identify twenty different amino acids? The answer is that the cell reads it three letters at a time – CAU, CAG, GGA, etc. This system has 64 (4 x 4 x 4) different combinations, or **codons**, which is more than enough to represent twenty amino acids. In fact, most amino acids can be represented by several alternative codons. And some codons are not instructions for amino acids at all, but signal the *end* of the gene (so-called "stop codons"). Cracking the genetic code was another one of Crick's insights, working with his Cambridge colleague **Sydney Brenner**. The complete set of codons was worked out in the 1960s, mainly from experiments using artificially made RNAs with just one or a few triplets in their sequence, crucial work begun by Marshall Nirenberg and Heinrich Matthaei, working at the US National Institute of Health laboratory in Bethesda, Maryland.

On the factory floor

To make a protein, a cell homes in on the corresponding piece of DNA code in the nucleus and makes an RNA copy – a single strand made up of As, Us, Cs and Gs. This is a story in itself, involving a large complex of proteins, most vitally the key enzyme called **RNA polymerase** (or RNApol). You'll have to refer to a molecular biology textboook for the details, but we now have a good idea about the operations of this extraordinarily intricate micro-machine that produces the vital flow of messages instructing which proteins must be made at any given time. RNApol constructs RNA chains in a process known as transcription.

In a eukaryotic cell, the newly made RNA is "trimmed" and often cut and spliced before it leaves the nucleus, to produce a piece of messenger RNA (mRNA). This splicing process was discovered in about 1977, when scientists realized that the genes of most living things are divided up into **exons** (*expressed* sequences), which are part of the instruction for the protein, and **introns** (*intervening* sequences), which are not. We still do not really understand why, although as we will see the splicing is one of the reasons why higher organisms can make many more different proteins than they have genes.

The trimmed mRNA molecule leaves the nucleus via tiny holes, and enters the cell cytoplasm, the "factory floor" where the real work begins. The mRNA is snatched up by the **ribosomes,** tiny protein-assembly machines (composed of ribosomal RNA and ribosomal proteins). Each ribosome works its way along the mRNA, reading the code from "start" to "stop", selecting the correct amino-acid building blocks and ejecting a growing protein.

RNA TRIPLET CODE FOR ALL TWENTY AMINO ACIDS

Ala	A	GCU, GCC, GCA, GCG	Leu	L	UUA, UUG, CUU, CUC, CUA, CUG
Arg	R	CGU, CGC, CGA, CGG, AGA, AGG	Lys	K	AAA, AAG
Asn	N	AAU, AAC	Met	M	AUG
Asp	D	GAU, GAC	Phe	F	UUU, UUC
Cys	C	UGU, UGC	Pro	P	CCU, CCC, CCA, CCG
Gln	Q	CAA, CAG	Ser	S	UCU, UCC, UCA, UCG, AGU,AGC
Glu	E	GAA, GAG	Thr	T	ACU, ACC, ACA, ACG
Gly	Gz	GGU, GGC, GGA, GGG	Trp	W	UGG
His	H	CAU, CAC	Tyr	Y	UAU, UAC
Ile	zI	AUU, AUC, AUA	Val	V	GUU, GUC, GUA, GUG
Start	AUG		Stop		UAG, UGA, UAA

Key:

Ala: Alanine	Cys: Cysteine	Asp: Aspartic acid	Glu: Glutamic acid
Phe: Phenylalanine	Gly: Glycine	His: Histidine	Ile: Isoleucine
Lys: Lysine	Leu: Leucine	Met: Methionine	Asn: Asparagine
Pro: Proline	Gln: Glutamine	Arg: Arginine	Ser: Serine
Thr: Threonine	Val: Valine	Trp: Tryptophan	Tyr: Tyrosine

The right amino acids are lined up using special "adaptors" – another discovery that was mainly due to an idea from Francis Crick that turned out to be right. The adaptors are themselves small RNAs called **transfer RNAs** (tRNA). Each has a three-letter **anticodon** corresponding to the three-letter codon in the messenger, and is attached to the corresponding amino acid.

To pick one example, the codon CGU in the mRNA has a complementary tRNA with the anticodon GCA, attached to an amino acid called **arginine**. Each tRNA brings with it the correct amino acid, which the ribosome then adds to the growing protein, releasing the tRNA at the same time. It takes much longer to explain this process than the cell takes to do it – a ribosome can select and add an amino acid in one 50th of a second. A cell can assemble a small protein like insulin in just a few seconds.

Finishing touches

As they fall off the **ribosomes**, the protein chains begin to fold up, taking on complex shapes. Chemical groups – small clusters of atoms joined together – interact with each other in the **protein chain** and with the water molecules around them in the cell; as the chain flexes and twists, they eventually find a unique folding pattern in which all the groups are in the environment they prefer. This fixes the protein's three-dimensional shape, which is what fits it for the role it plays in the cell – whether it is a piece of "scaffolding" (a **structural protein** like keratin), or an **enzyme** which catalyzes reactions between specific molecules. The key point is that (double-stranded) DNA always has the same overall shape, whatever its sequence. But proteins, on the other hand, are always different.

Many proteins then need to combine with other proteins before they can do their job, while others need to have fats or carbohydrates attached. Some proteins are first made as an inactive form, and later chopped into a smaller, active form. For example, the enzymes – the biological catalysts that control the rate of chemical reactions in your body – are initially inactive, and are then activated as and when they are needed. Once made, some proteins stay in the cytoplasm. Others move to the cell nucleus or cell membrane (the fatty outer layer of the cell), in order to be used. Still others are exported outside the cell, to carry out various tasks in the body. The hormone **insulin**, for example, leaves the pancreas cells in which it is made and enters the bloodstream to adjust sugar levels.

Changes in the DNA

Rewriting the code

Sometimes, the DNA code of a living creature can acquire changes, called **mutations**. Mutations can either be harmful, beneficial, or of no consequence – it depends on whether they affect a protein or not, and how they affect that protein. One thing's for sure, without them we'd all still be floating around in the primeval soup, because genetic mutations are the driving force behind natural selection (see chapter 5 for more on evolution). But mutations can also cause disease, by adversely affecting the way the body grows or functions.

Although the entire human genetic code is made up of nearly three billion base-pairs of DNA, most of it (over 98 percent) doesn't contain any genes. Returning to the factory archive analogy, it's the equivalent of reams of boring paperwork stuffed in the filing cabinets, in between

DNA and the origins of life

DNA needs proteins to read the information it contains and help run the cell. But the manufacture of proteins itself depends on DNA, which contains the information about what order to stitch amino acids together. It's a chicken-and-egg puzzle, but it's one that is a key part of the problem of understanding how life began all those billions of years ago. Many biologists now believe that the answer involves **RNA**.

In living things as we see them now, the "other" nucleic acid combines the properties of DNA and protein. Like DNA, it can preserve information in the order of bases in a linear molecule. Like protein, single-stranded RNA can fold up into unique shapes, as short stretches of bases on the single strand pair up and create loops and hairpins in the molecule.

This means an RNA molecule can be used either as a message or as an enzyme. As the chemical building blocks of the nucleic acids probably formed fairly easily in the environment of the young Earth, it is possible to imagine an "RNA world", in which proto-organisms had lots of different RNAs reproducing. Eventually, some arose that helped to catalyze first their own reproduction, then perhaps became specialized enough to act as enzymes that promoted reproduction in general. The division of labour between the other chemical stars of the cell – DNA and protein, with RNA mediating between them – would have come later.

We cannot be sure what happened, in what order, so long ago. But it may soon be possible to reproduce the key steps – and show what *could* have happened – in the lab.

the all-important designs. This "non-coding" DNA is sometimes called **junk DNA**, although this term is becoming less popular, as scientists are realizing that much of it does have a purpose. The fact remains that the odd change here and there in a stretch of non-coding DNA probably won't have much effect. But occasionally a mutation occurs in a gene that changes the RNA molecule it gives rise to, and sometimes affects the protein it codes for.

Since the **RNA code** (like DNA) uses four chemical "letters" in groups of three, changing just one letter can have a drastic effect on a gene – just as changing one letter in a sentence of three-letter words can alter its meaning. For example, changing one letter in the sentence "THE CAT WAS FED" could give you "THE CAR WAS FED" or "THE CAT WAS RED". In the same way, a single "letter change" in a piece of DNA code can change the resulting mRNA, so that a different amino acid is inserted at that point in the protein chain. For example, the DNA code CCG GAC ACG AAA tells the cell to add the amino acids **proline**, **asparginine**, **threonine** and **lysine** to the protein it is making. If just one of the DNA bases is changed, a different amino acid may be substituted.

Not all genetic mutations involve changing a single letter of DNA code – inserting or deleting a letter can have equally, if not more, dramatic effects. Inserting a random letter into "THE CAT WAS FED" sentence would render it meaningless: "THE HCA TWA SFE D", as would removing a letter. Because the RNA code is also read three letters at a time, inserting or deleting a single base can mess up the whole protein. Other types of mutation can involve much bigger deletions or insertions, affecting one or more whole genes. Sometimes, sections of DNA are duplicated, or repeated many times in a kind of "genetic stutter".

Mutations and disease

Most DNA mutations occur when your cells divide to make two new cells, as this entails copying the entire genetic code. With 3,000,000,000 bases to copy, a few mistakes creep in every time. DNA is under constant attack from other chemicals in cells, which can also cause mutations. These can be generated from the normal workings of everyday cell life. The energy-generating machinery of the cell, for example, tends to release highly reactive **free radicals** which can react chemically with bases in DNA. And some chemicals from outside, such as components of cigarette smoke, can damage DNA too.

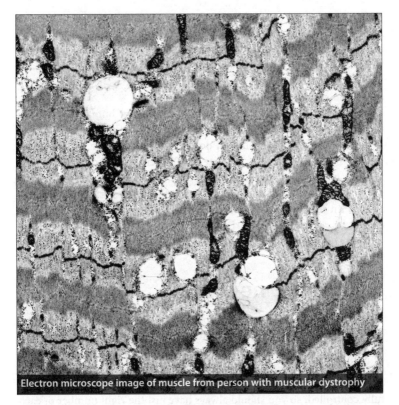

Electron microscope image of muscle from person with muscular dystrophy

Fortunately, cells have a very efficient DNA repair system that is constantly scanning DNA. It works partly by looking for mismatched bases in the double helix, which tend to distort its structure. Without these DNA repair enzymes, cells would quickly accumulate too much damage to survive. But like every system it has its flaws, and DNA mutations occasionally slip through the net. As explained above, how significant this is depends on whether the mutation affects a protein or not. Even then, it depends on the cell in question and what its purpose is: for example, if a hair cell acquires a mutation in a blood protein gene, it won't matter, as a hair cell will never need to make blood proteins anyway.

There are two places where mutations can have a potentially devastating effect, however. The first is in the genetic code of egg and sperm cells, as these mutations can be passed on to the next generation. As well as driving evolution, these changes can also lead to problems. A mutation that produces a faulty protein could result in an embryo that can't develop properly, or a baby being born with a genetic condition such as **cystic fibrosis** or **muscular dystrophy**. The other instance in which mutations can cause serious health problems is when they occur in genes that control cell growth. A cell with a faulty growth-control protein can potentially start to divide and multiply in an uncontrolled way – in other words, it can become a **cancer cell**.

These major implications of genetic mutations for human health were two of the main reasons behind the recently completed **Human Genome Project** (HGP): the multi-billion dollar effort to read the entire human genetic code. Hatched in the late 1980s, scientists thought that this massive undertaking would be the best way to understand the genetic mutations that cause cancer. Furthermore, the US Department of Energy needed an efficient way to study possible inherited genetic mutations caused by radiation – in particular, from the atom bombs dropped on Japan in 1945. (For more on the HGP, see chapter four.)

New cells, new DNA

Your body is constantly replacing old cells with new ones, at the rate of millions per second. In fact, by the time you finish reading this sentence, 50 million of your cells will have died and been replaced by others. Some are lost through "wear and tear", some just reach the end of their life, and others deliberately self-destruct. The life cycle of every cell is carefully controlled, so you should always have just the right number of each type of cell. The stages of a cell's life are called the **cell cycle**. This works a bit like a washing machine cycle, in that each stage must be finished completely before the next one is allowed to start. Instead of wash, rinse, spin and drain, the cell cycle is made up of stages in which the cell grows and rests, copies its DNA, and divides into two new cells. Because this is a book about genes and cloning, this section will only deal with how cells copy their DNA.

Before a cell can divide, it must unravel its chromosomes and copy all of its DNA, so that each new cell gets a complete copy of the genetic material. On average, this takes seven hours, after which the cell divides into two new ones – a process called **mitosis**. Cells that are making egg or sperm cells must divide in a different way. Each egg and sperm cell ends up with only half the amount of genetic material present in the original cell, bundled up as 23 chromosomes. This special way of dividing is called

meiosis, and means that when egg and sperm meet at fertilization, the resulting embryo will have 46 chromosomes.

In the process of mitosis, the DNA that makes up the 46 chromosomes first "unzips", with the help of special enzymes, allowing it to be copied. The chromosomes – usually long strings in the nucleus – then coil up into compact bundles. The edge of the nucleus dissolves, and a temporary protein scaffolding appears in the cell: the **spindle apparatus**. The chromosomes then line up in the centre of the cell, attaching themselves to the spindle apparatus for support. The duplicated chromosomes then split up, and one of each moves to the opposite end of the cell. Finally, the cell narrows in the middle and divides.

Meiosis is a little different. The 46 chromosomes in a human cell actually consist of 23 pairs: numbers 1–22 plus the sex chromosomes, X and Y (men have one of each sex chromosome, while women have a pair of X-chromosomes). When cells divide to make egg or sperm cells they need to receive only one chromosome from each of the pairs – one number 1, one number 2, and so on. To achieve this, cells that divide by meiosis actually divide twice. First, the cell copies the DNA in all 46 chromosomes – a complete copied set of all 23 pairs is made. Then, the chromosomes line up in the centre of the cell, just as in mitosis. But unlike in mitosis, the chromosomes line up in their pairs, rather than as 46 individual chromosomes. Then one chromosome from each pair moves to each end of the cell. The cell then splits into two: each new cell with 23 chromosomes. These new cells divide again, but this time each duplicated chromosome splits down the middle (as they do in mitosis). The result: four cells, each with 23 chromosomes.

As any parent knows, children with the same mother and father can look completely different (unless they are identical twins). This is mainly down to some genetic shuffling around that goes on when a cell divides to make eggs or sperm. During meiosis, the two chromosomes in each pair become closely entwined around each other and swap parts of themselves. This process, called **recombination**, shuffles the genetic information – two stretches of DNA that were near each other on one chromosome may end up in *different* egg or sperm cells. Recombination ensures that every individual egg and sperm cell contains a unique set of genetic information. Occasionally, meiosis goes wrong. The physical and cognitive impairments associated with **Down's syndrome** are an example of this. This condition is usually the result of an error during meiosis in which an egg or sperm cell receives an extra chromosome 21.

Controlling genes

The same, but different

How do different cells manage to make completely different proteins, despite having identical genetic information? The answer is that any particular cell uses only a *selected* set of genes – just as a chef might select only a few recipes from his or her repertoire to put on the lunch menu. The other genes remain "switched off", so no proteins will be made from them. To "switch on" a gene, key proteins called **RNA polymerases** (see p.22) have to latch onto the start of the gene's DNA code, and begin to take the RNA copy needed to make the protein. But RNA polymerases can't do this by themselves – they need other proteins, known as **transcription factors**, which stick to the start of the gene and get things going. These transcription factors – and the bits of DNA code they stick to – vary between genes, providing a way in which different cells can switch on the subsets of genes they require. In fact, these control mechanisms can be very precise: gene switches are more like the dimmer switches sometimes used for electric lights, rather than the simple on–off variety.

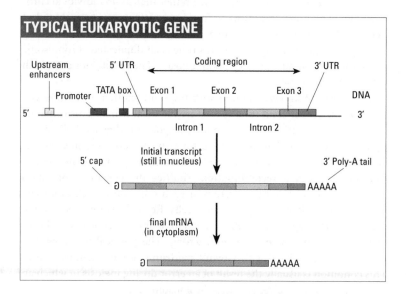

TYPICAL EUKARYOTIC GENE

As well as switching on genes, other transcription factors can switch off genes, by sticking onto different bits of DNA code associated with the gene. The region of the gene that contains all the DNA code that the various different transcription factors might alight upon is called the **promoter region**. And transcription factors are themselves coded for by genes, with their own promoters, so complex control networks and cascades can evolve across the genome.

Nor is this the only level at which genes can be controlled – the activity of a gene also depends upon how tightly packaged it is in the chromosome, and whether certain chemicals have been attached to its DNA. In particular, if chemicals called **methyl groups** are attached to a gene (specifically to the "C"s of its DNA), it generally means it's not doing much, and will remain inactive in the cell. The study of how genes can be switched off and on – sometimes in a way that can be passed on from one generation to the next – is known as **epigenetics**. This buzzword looks increasingly set to take centre stage, as geneticists move on from identifying genes to figuring out how they exert their influence.

Genes, people and families

The first chapter provided a whistle-stop tour of the inside of a cell, along with a brief outline of how cells store, use and copy genetic information. The rest of this book will mainly concern itself with the implications of this basic knowledge – in terms of our health, culture, society and future. But first, we will look at how genetic information orchestrates the growth of a new person, and how genes are inherited.

From conception to birth

We all began life as a single cell: a fertilized egg that received half its genetic information from its father's sperm, and half from its mother's egg. The vast majority of this DNA came from the nucleus of each of these two cells, which is united in the nucleus of the new cell – officially called the **zygote**. However, a tiny bit of our genetic material comes from the egg cell's **mitochondria**, which are inherited solely from the mother (see box opposite). The zygote is only known as such until it starts to divide, making first two cells, then four, then eight, and so on – doubling again and again into billions of cells that together form the **embryo**. The embryonic stage of development – in which the basic body shape is laid down, and all the major organs are established – lasts for nine weeks in humans, after which it is known as a foetus. (This stage is actually dated as being the eleventh week of pregnancy, as doctors calculate the estimated date of delivery of the baby from the beginning of the mother's last menstrual period – on average, two weeks before fertilization.)

During weeks 9 to 38 the foetus gets bigger and its organs continue to grow and develop. This "post-embryonic" stage of development continues well after birth, which occurs at around 38 weeks after fertilization (the 40th week of pregnancy). The brain of a newborn baby, for example, is not fully formed until about the age of two. This is the result of a "trade-off" that humans evolved, which enabled us to keep both our big brains

Mitochondria and your mum

All your body's cells contain mitochondria, tiny sausage-shaped powerhouses. Mitochondria have their own small number of genes, which make some of the proteins needed to generate energy for the cell. All your mitochondria are copies of the 100,000 or so that were inside your mother's egg cell. Your father's sperm also contained mitochondria, but these were all packed inside the "tail", which breaks off and is lost after the sperm enters the egg. So mitochondria are passed from a mother to her sons and daughters, but only her daughters will pass them on to their own children. Most body cells have around 500–1000 mitochondria, so there are many copies of the mitochondrial genetic material in each cell – a short piece of mitochondrial DNA is often the only sort present in sufficient quantities to be detected in ancient remains. Scientists have used this fact – along with the exclusively maternal inheritance of mitochondrial DNA – to track the migration patterns of our ancient (female) ancestors.

Mitochondria are about the size and shape of bacteria, probably because this is exactly what they once were. Scientists believe that, very early on in the history of life on earth, bacteria-like creatures that could use oxygen to make energy were engulfed by other simple cells, forming a mutually beneficial relationship. The cells that went on to become mitochondria received free nutrients, and in return, the host cell got energy to use for its own chemical reactions. (Chloroplasts – the structures in plant cells that capture light energy – probably arose in a similar way.) This idea – a phenomenon termed **endosymbiosis** – was widely dismissed when first proposed by US scientist Lynn Margulis in 1966, but there is now good evidence to support her theory of cell evolution. For example, mitochondria and chloroplasts both have a single, circular piece of DNA, just like modern-day bacteria.

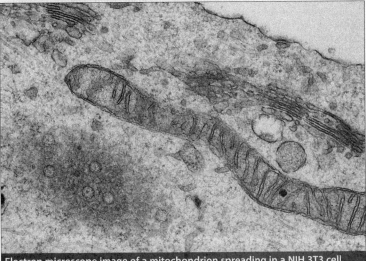

Electron microscope image of a mitochondrion spreading in a NIH 3T3 cell

Animals, embryos and Hox genes

The animals used to study embryo development include the fruit fly *Drosophila melanogaster* and the **nematode worm** *Caenorhabditis elegans* (an animal that always ends up with exactly 959 cells is a gift to scientists studying the fate of cells during embryo development). Advantages common to both flies and worms, for the purposes of experimentation, are that they can be bred in large numbers in the lab, and have transparent embryos. Other favourites of developmental biologists include the domestic chicken *Gallus gallus* and the African clawed frog *Xenopus laevis* – both of which have embryos that develop entirely outside the body. The ubiquitous lab mouse, *Mus musculus*, has also been much studied, being a close relative of humans, and so has the versatile zebrafish (*Danio reiro*), which is both easy to breed in large numbers and has easily accessible embryos.

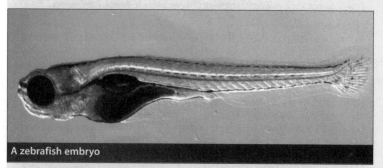

A zebrafish embryo

Scientists have found that, quite unexpectedly, many of the genes that control animal embryo development have remained virtually unchanged throughout evolution. When it comes to laying down the animal body plan, it seems that nature is not really interested in reinventing the wheel. Versions of the genes that specify where the body parts should go do pretty much the same job in a fly embryo as they do for a mouse or human embryo. These **Hox** (short for **homeobox**, which refers to a crucial section of DNA code shared by all members of this gene family) genes were discovered by American scientists **Mike**

and our upright posture: if a human baby stayed in the womb until it was fully developed, then its head would be too big to fit through the woman's relatively narrow pelvis. It is during the early weeks of pregnancy when the most dramatic developmental changes occur, involving many complex processes that we are only gradually beginning to understand. You can find out more about what happens during each week of a human pregnancy at **The Visible Embryo** website (www.visembryo.com).

Levine and **Bill McGinnis** in 1984, when they were studying fruit fly embryos. They found that mutations in Hox genes can have drastic results – producing flies with feet in place of antennae (antennapedia mutants), or an extra set of wings for example. Researchers soon realized that all animals have Hox genes, and that they have probably played a key role in evolution, since even a slight change in a single Hox gene can have a dramatic effect on an animal's body. Snakes may have evolved from lizards by losing the Hox gene for making legs. Occasionally, long-gone features (known as **atavisms**) can reappear in animals – such as whales and snakes with basic limbs, horses with hooves split into three "toes", and even people with small tails. These are all probably caused by changes to Hox genes that resurrect ancient body plans.

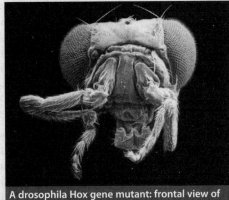

A drosophila Hox gene mutant: frontal view of antennapedia head

Many of the genes involved in embryo development control several other genes, which in turn switch on others – resulting in self-contained "modules" of biological information. For example, a mouse gene that controls eye development can, if transferred into a fruit fly embryo, form fly eyes wherever the gene is switched on. Although there are many genes involved in the growth of an eye, control genes such as this one trigger a localized cascade of gene activity. Changes in these genes provide a mechanism for "evolutionary short cuts", in which whole sections of the body can potentially appear, disappear or change locations. The study of these controls and their history, under the banner of evolutionary developmental biology (or "**evo devo**") for short, is a very active area of current research.

Genes and the growing embryo

Each cell in the growing embryo must end up in the right place, doing the right job. To do this, the cells have to proliferate, move around, specialize, rearrange themselves and sometimes even commit suicide. The fate of a particular cell – whether it becomes a toe, liver or ear cell – depends on its location, the instructions it receives from other cells and those it receives from its own genes. Each new cell in the growing embryo receives two full sets of around 25,000 different genes (one from each parent). But not every cell needs to make every protein, and cells make some proteins only at certain times. As outlined in the first chapter, gene activity is precisely controlled via **transcription factors** – proteins that either crank up or damp down the production of a particular protein (see p.30). The genes that make these proteins are important in every living cell, but especially so in the cells of the developing embryo.

Unravelling the complex series of events that control human development is a major focus of biomedical research, which involves the study of many different genes, the proteins they make and their interactions with other genes and proteins. Our knowledge is still incomplete since studying human embryos is difficult for both ethical and practical reasons. Scientists have therefore taken the less direct but highly successful route of identifying **animal genes** important in embryo development, and then tracking down their human counterparts.

It's important to stress that genes are not the only thing that can influence the developing embryo and foetus. It is highly sensitive to its environment in the womb. The mother's health and diet during pregnancy are crucial to foetal development, with recent research suggesting that these effects can last well into adulthood. The placenta provides vital nutrients, but also a route by which other potentially harmful substances can reach the embryo or foetus. For example, thousands of babies worldwide were born with birth defects after pregnant women in the 1950s and 1960s were given the drug **thalidomide**, with the aim of stopping severe morning sickness. Substances that interfere with embryo development and cause congenital problems such as this are called **teratogens** – other examples include the rubella (German measles) virus and alcohol, when excessively consumed.

Embryo development: cell specialization

In the first two days of life, the newly fertilized human egg travels along the mother's Fallopian tube to the womb. On its way, it divides to make a clump of 32 cells – this is called the **morula** stage. If the early embryo splits into two clumps of cells before this stage, it can grow into identical twins. This is because every one of the cells in the morula could still become any part of the growing embryo (non-identical twins arise when two sperm fertilize two separate eggs). As the cells of the embryo continue to divide and multiply, they begin to specialize, creating increasing levels of complexity. In their definitive textbook *Human Molecular Genetics*, geneticists **Tom Strachan** and **Andrew Read** use an elegant analogy for embryo development. Imagine a stream running down the side of a mountain, they suggest. Individual cells are like leaves carried by the water. The stream may branch many times before it reaches the bottom of the mountain, but each leaf only follows one path. Every time it approaches a different branch, a leaf must therefore commit to one path or the other, an action that is irreversible. The cells that make up the very early embryo are described as being **totipotent**: they still have the ability to follow any pathway in the developing embryo, analogous to the leaves at the very top of the mountain. As development proceeds, the cells become increasingly more specialized, until they end up as part of a particular body tissue.

At six days old, the embryo, now called a **blastocyst**, still consists of only a few hundred cells. But these cells have already started to organize themselves – those on the inside will eventually grow into the foetus, and those on the outside into the placenta. Over the next couple of days, the embryo burrows into the lining of the womb, a process known as **implantation**. This event triggers the production of the hormone detected in a pregnancy test, called **human chorionic gonadotropin** (hCG). Inside the blastocyst is a clump of cells termed the inner cell mass, which will eventually give rise to all the different organs and tissues of the body. Because of their unique ability to develop into any type of body cell, these so-called **embryonic stem (ES) cells** are of great interest to scientists working to understand and develop new treatments for many incurable conditions. But while stem cell therapies have shown great promise, they have also sparked considerable controversy, especially in the US (see chapter 12 for more on this debate).

Which way up?

The cells closest to the placenta are destined to become the front of the body, and the top and bottom ends of the embryo are also defined very early on. Although this process is quite well understood in several animal species, how human embryos know which way up they are remains something of a mystery. In chickens, the head-to-tail orientation is defined by gravity, as the fertilized egg makes it way down the **oviduct** (egg-laying tube). In frogs, each end of the unfertilized egg is distinguished by the presence of different proteins, with the sperm entry point providing another point of reference. It's likely that in mammals too, the site of sperm entry provides some vital orientation information. However it is achieved, once established, the specific "head" and "tail" genes are switched on at either end of the human embryo – ensuring that the right body parts end up growing in the right place. The other axis laid down in the early embryo is the left–right **asymmetry** of the body: although the human body looks roughly symmetrical from the outside, much of the inside is asymmetrical. For example, your heart is towards the left side of your body, and the liver to the right, while the right kidney is positioned lower than the left kidney.

Laying down the body plan

Another, related event comes in the third week of human development. The mass of cells form three different layers – known as the **ectoderm**, **endoderm** and **mesoderm**. This whole process, along with the emergence of the embryonic axes, is known as **gastrulation**. It involves massive movements of cells, and is a crucial step on the path to **differentiation**, the gradual emergence of distinct types of tissue, expressing different combinations of genes. Between them, these three layers of cell will form all of the two hundred or so different tissues of the embryo. The outer layer (ectoderm) grows into skin, brain and nerves. The middle layer (mesoderm) grows into muscle, blood vessels, bones and many of the organs. The inner layer (endoderm) grows into the gut, stomach and lungs. Just before this stage, the embryo looks like a flat disc, with a groove called the **primitive streak** running through the middle. To start shaping the body, a temporary scaffolding grows in the centre of the embryo (the **notochord**). Above it, a structure forms called the **neural plate**, which folds up to form the neural tube – imagine a paper plate being rolled up. Occasionally, the neural tube – which will grow into the brain and spinal cord – doesn't completely close, resulting in conditions such as spina bifida. A lack of folic acid in the mother's diet before and during early pregnancy is known to increase the risk of spina bifida, but genetic factors are also involved.

Fleshing out the details

Like an artist developing a basic sketch, once the body plan has been laid down, the details can start to be filled in. The early embryo is next divided up into blocks called **somites** – each of which is under the control of so-called Hox genes (see box on p.35). The somites will later form the body and limb muscles, ribs and backbone. Which tissues are made in which somite is dependent on its position in the embryo, which in turn is influenced by the proteins made by the Hox genes. These chemical signals are switched on at different points along the length of the embryo, providing a sort of "grid reference" for each cell.

Limbs grow between the fourth and eighth weeks of development, starting off as tiny bumps called **limb buds**. Cells are told which part of the limb to form by a combination of chemical gradients flowing along each physical axis. In the arm, for example, these would be from shoulder to hand, from left to right and from front to back – so that each cell gets a three-dimensional grid reference of its location.

The hands and feet of human embryos start out as webbed – to make fingers and toes, rows of cells have to neatly die, via a controlled process known as **apoptosis** (see box below). Organs also form during this time, as layers of unspecialized cells are coaxed down different developmental pathways. To do this, cells need to turn up or down production of proteins from particular genes. They also need to communicate with each other – when one cell receives the instruction "make the proteins you need to become a liver cell", it passes the message onto its immediate neighbours.

Apoptosis: a good death

A cell can die in many ways – through infection, poisoning, overheating or lack of oxygen. An uncontrolled cell death is messy: the cell swells up, and its contents leak away, perhaps damaging surrounding cells. But there is another, tidier way to go: programmed self-destruction, or apoptosis. The term was coined in a seminal 1972 article, but scientists had been aware of "programmed cell death" for some time. The last fifteen years or so have seen a huge amount of research into this crucial biological process, which is vital for normal human development and good health throughout life. As well as helping to shape the growing embryo, apoptosis prunes unwanted nerve cells in the developing brain; makes sure that organs stay the right size in adult bodies; gets rid of cells with damaged DNA; and weeds out ineffective immune system cells. The process is not restricted to humans – apoptosis is vital to every developing animal embryo, and is the driving force behind the spectacular transformations of caterpillars into butterflies and tadpoles into frogs. Some plant cells also deliberately kill themselves to stop an invading fungal or bacterial infection spreading.

To kill itself, a cell first switches on the genes that together make a deadly chemical cocktail. It then separates itself from its neighbours, and unleashes the poisons. These include a substance that chews up the DNA in the cell nucleus and a "glue" that binds the inside of the cell together. Within a few hours, the cell shrinks, breaks up and is engulfed by other cells. Increasingly, scientists are realizing that disruptions to apoptosis could be the key to understanding some illnesses. A failure of genetically damaged cells to kill themselves can result in cancer – an unchecked cell growth. At the other extreme, excessive apoptosis plays a role in wasting disorders such as Alzheimer's and Parkinson's disease.

The male gene switch

The sex of a human embryo is determined at conception. Women inherit two X-chromosomes, while men inherit one X and one Y-chromosome. Eggs and sperm have only one sex chromosome each; all eggs have one X-chromosome, while half a man's sperm will have an X, and the other half a Y-chromosome. So if a Y-carrying sperm fertilizes an egg, the resulting embryo will be male, whereas a sperm bearing an X-chromosome will produce a female embryo. However, in a seven-week-old embryo, the sex glands and organs of males and females appear identical. At this point a "male gene switch" is flipped in male embryos. The testis determining factor gene, called **SRY**, is found on the Y-chromosome. It triggers male development in all mammals, and causes the testes and male external genitals to start to grow. In a female embryo, there is no Y-chromosome and so no SRY gene, allowing the ovaries and female external genitals to develop – in mammals, being female is therefore the 'default' state for all embryos.

Genes, families and inheritance

So much for how genes orchestrate the growth of a new person. What about the effects they have after birth? Genes affect the way you look, your health, and the way your body works. You may have been told that you have your granddad's nose, or your auntie's eyes. You may also already have, or be predisposed to, certain health problems due to genes passed on by your parents. Even your personality and behaviour are to some extent influenced by your genetic make-up, though this is an area of research that has proven to be extremely controversial (see chapter eight).

To study the effects of different human genes, geneticists study families, and the similarities and differences between related and unrelated individuals. But the basic principles of heredity were originally figured out using **pea plants**, by Gregor Mendel, a nineteenth-century monk.

Mendel and his laws

No book about genetics and cloning would be complete without an overview of the work carried out by Mendel, an Austrian dubbed (many years after his death) "the father of genetics". After failing to qualify as a teacher, he spent his life studying peas and recording the characteristics of thousands of plants grown in the monastery gardens in Brunn (now Brno, Czech Republic). Mendel lived from 1822–1884, many years before genes, DNA and chromosomes were discovered. He studied the *effects* of genes (which he called **factors**), by tracking the inheritance of different traits from one generation of plants to the next. As happens to many big ideas in science, his theories and observations were largely ignored at the time, only to be rediscovered years later – some forty years later in Mendel's case.

Gregor Mendel

To simplify matters, Mendel confined his investigations to traits that appeared in one of just two forms, such as seed colour (yellow or green), seed shape (smooth or wrinkled), flower colour (purple or white) and height (tall or short). By cutting off the pollen-laden stamens to prevent self-fertilization, and covering the flowers with cloth bags, Mendel was able to control which peas were fertilized in each of his experiments. First, he spent two years producing pea plants that "bred true" for each trait – that is, tall plants which always produced tall offspring, and so on. He then set about meticulously crossing different varieties, growing an estimated 28,000 pea plants during his studies.

Mendel discovered that when he crossed two different varieties of true-breeding peas, only one of the two traits appeared in the next generation. Crossing a short with a tall plant, for example, always produced tall plants, never peas of short or medium height. When Mendel crossed a plant with yellow seeds and a plant with green seeds, all the offspring had yellow seeds. He concluded that one of the two traits in each pair was always dominant over the other, and he used the term "recessive" to describe the trait that did not appear in the next generation. The words **dominant** and **recessive** are still used today to describe inherited traits.

Further experiments showed that recessive traits masked in one generation of pea plants could resurface in the next: when Mendel bred two of the tall pea plants obtained from the crossing of a tall and a short plant,

Mendelese: some key definitions

What Mendel called factors, we now call genes. Different versions of the same gene, such as those that give rise to either tall or short plants, are known as **alleles** (pronounced a-LEELs). Plants that "breed true" do so because they have two identical alleles, and are known as **homozygotes**, whereas plants with two different alleles are described as **heterozygotes**. These definitions apply to people and animals, too. An individual's outward appearance with regard to a trait, be they human, pea plant or mouse, is called its **phenotype**, whereas its genetic constitution is referred to as its **genotype**. For example, a tall plant may have one short and one tall allele (its genotype), but because the tall allele is dominant, its appearance (phenotype) is tall.

he found that about three quarters of the new batch were tall, but about one quarter were short. After carrying out many experiments of this kind, Mendel surmised that each plant had two copies of the factor influencing a particular trait, which never blended together. In the next generation, each plant passed on just one of the two factors at random to each of its offspring, a principle which he called the **rule of segregation**.

Mendel then set about tracking the inheritance of more than one trait, for example by crossing true-breeding short, smooth-seed plants with tall, wrinkled-seed plants. In the first generation, all the offspring would be tall with smooth seeds, because they are the dominant forms of each trait. But in the next generation, he might end up with any combination of plant – tall with smooth seeds, short with smooth seeds, tall with wrinkled seeds or short with wrinkled seeds. He discovered that the two traits behaved completely independently of each other. He called this the **rule of independent assortment**. We now know that Mendel was incredibly lucky with the characteristics he chose to study, since inherited traits only behave in this way if the genes responsible are located on different chromosomes. Also, not all traits have clear-cut dominant and recessive forms, and there are several other types of inheritance patterns. Despite these caveats, Mendel's ideas have stood the test of time, and are still taught to every high-school biology student – not a bad legacy for a failed teacher. You can read more about Mendel in Robin Marantz Henig's occasionally fanciful *A Monk And Two Peas: The Story Of Gregor Mendel And The Discovery Of Genetics.*

Genes and families

Your genetic material is made up of 46 chromosomes, which contain two sets of around 25,000 different genes – one set inherited from each of your parents. The chromosomes come in pairs: 44 **autosomes** numbered 1–22 (almost in order of decreasing size, except that chromosome 21 is smaller than 22), plus a pair of sex chromosomes (XX or XY). So you inherit your genes in pairs too, with one of each pair coming from each parent. Most of these genes come in several alternative varieties, called alleles. Although you may share a particular allele with many other people, no one in the

The Hapsburg lip (and chin), manifest upon King Philip IV of Spain – its most celebrated bearer, thanks to Diego Velazquez.

Studying human inherited traits

Geneticists studying people cannot, of course, carry out controlled breeding experiments to work out how different traits are inherited, as their colleagues working on peas, flies, mice or worms can. But they can gain a lot of information by studying families, and taking detailed family history information. Methodically tracking down family members and obtaining DNA samples from them is the crucial first step in any gene hunt aiming to find whether a single gene might be responsible for an inherited disease. The faulty genes that cause cystic fibrosis, Huntington's disease, Fragile X syndrome and many other genetic conditions have all been identified in this way – most during the human gene "gold rush" of the late 1980s and 90s. To track the inheritance of a disorder, or indeed any characteristic, geneticists can record family history information using pedigree diagrams – a special sort of family tree (see p.46).

The inheritance patterns of traits caused by single genes (so-called **monogenic traits**) are relatively easy to work out. However, as we've said, most human characteristics are influenced by several different genes (**polygenic traits**). Furthermore, traits like height, as well as most common diseases, are the result of many genetic and non-genetic factors – such as diet, lifestyle and womb environment during early development. Working out which genes are involved in "multifactorial diseases" is much more of a challenge, and is likely to keep geneticists busy for at least another couple of decades. Rather than studying individual families, it requires large-scale population studies, and detailed comparisons of the genes of affected and unaffected individuals (see chapter four for more on these ongoing efforts).

world has exactly the same *combination* of alleles as you – unless of course you have an identical twin. Sometimes, as happens in peas, the effects of one allele can mask that of the other. For example, someone with brown eyes may also have the instruction for blue eyes, but the effect of the brown-eye gene masks the effect of the blue-eye gene. The effect of genes masked in one generation can sometimes show up in the next. The vast majority of inherited human traits are influenced by many pairs of genes, so studying human inheritance is much harder than studying peas. However, Mendel's concepts of dominant and recessive genes do apply to many genetic conditions, and there are also a few non-disease examples.

The ability to taste the chemical **phenylthiocarbamide** (PTC) is a genetically determined, dominant trait; the lack of this ability being a recessive one. US researcher **Un-kyung Kim** and colleagues finally identified the gene responsible in 2003, and showed that it actually comes in three different varieties. It codes for a **taste receptor**, a protein found in the taste buds of the tongue. Depending on which two of the three gene versions they have, people find PTC either tasteless, slightly bitter, or very bitter. Another example of a simply inherited human trait is the "Hapsburg lip", a common feature in many members of the European royal families in the sixteenth and seventeenth centuries. This distinctive jutting-out lower jaw and lip is a dominant trait, and it can be passed on to a child even if only one parent has it.

Genetic conditions and inheritance patterns

People born with a simple genetic condition have a single inherited change in their genetic information (called a **mutation**), which in turn affects one or more of the proteins made by their bodies. Sometimes this means that the protein is still made, but there is too little or too much of it. In other cases, the protein is missing completely – for example, in people with the condition **Duchenne muscular dystrophy**, a crucial muscle protein is missing. In others, such as the inherited condition Huntington's disease, a "toxic" protein is made instead of the normal protein. There are over 5000 different genetic conditions, which together affect around three to five percent of all babies born. Sometimes, the illness clearly runs in the family, and several members are affected. But often an affected baby is born to unaffected parents, neither of whom have a family history of the disease. This can happen either because the genetic change responsible arose for the first time in the egg or sperm, or because one or both of the parents are healthy "**carriers**" of the disorder. Many genetic conditions have distinctive inheritance patterns, which can become apparent by recording information on which family members are affected. This type of analysis is usually the first step taken when genetic counselling is offered to a family affected by a genetic disorder.

Dominant inheritance

Huntington's disease (HD), polycystic kidney disease, achondroplasia and neurofibromatosis are all examples of genetic conditions that show an **autosomal** dominant pattern of inheritance – autosomal because the gene involved is located on one of the autosomes, rather than either of the sex chromosomes, and dominant because the effects of a faulty gene are *not* masked by a working copy. If one of the two parents is affected by a genetic condition with this type of inheritance pattern, every child has a one-in-two chance of being affected. So *on average* half their children will be affected and half their children will be unaffected. Another way of explaining a dominant disorder is to say that **heterozygotes** (people who

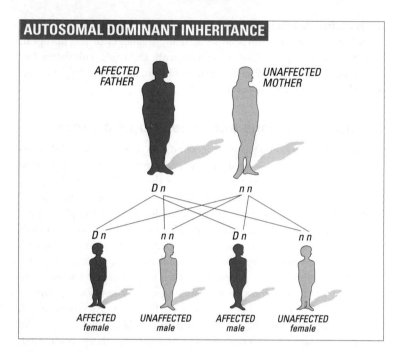

AUTOSOMAL DOMINANT INHERITANCE

AFFECTED FATHER

UNAFFECTED MOTHER

D n

n n

D n

n n

D n

n n

AFFECTED
female

UNAFFECTED
male

AFFECTED
male

UNAFFECTED
female

inherit one faulty and one working copy of the gene) and **homozygotes** (people who inherit two copies of the faulty gene) are both affected by the condition.

Recessive inheritance

Cystic fibrosis, sickle cell disease and Tay-Sachs disease all have an autosomal recessive pattern of inheritance – autosomal because the gene involved is found on one of the autosomes, and recessive because the effects of a faulty gene are masked by the effects of a working copy. Children with a recessive disorder are usually born to two healthy "carrier" parents, each of whom has one copy of the faulty gene. They have a one-in-four chance of passing on the disorder to any baby they have, so *on average*, one quarter of their children will be affected. There is also a one-in-two chance that each child will be a healthy carrier. Another way of explaining a recessive disorder is to say that **heterozygotes** (people who inherit one faulty and one working copy of the gene) are unaffected by the condition, while **homozygotes** (people who inherit two copies of the faulty gene) are affected.

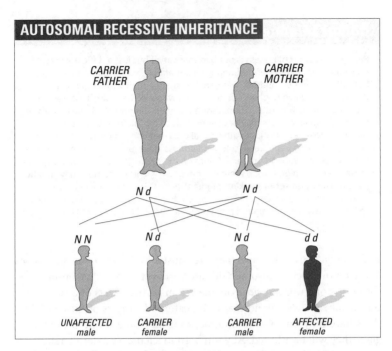

AUTOSOMAL RECESSIVE INHERITANCE

CARRIER
FATHER

CARRIER
MOTHER

N d

N d

N N

N d

N d

d d

UNAFFECTED
male

CARRIER
female

CARRIER
male

AFFECTED
female

Working out the consequences of inheriting a double dose of recessive gene alteration is often complex, as large genes may often have more than one possible mutation. The gene associated with cystic fibrosis, for example, has more than 1500 recorded mutations. Although some are much more common than others, two carrier parents may well have different mutations. If any of their offspring inherit both mutations (the chances of this being one in four, remember), their precise symptoms, and the course of their disease, will depend on the combination of altered alleles they possess. So although we often read about "simple" Mendelian inheritance of some diseases, even the simple cases seem to get more complicated as we learn more about them.

X-linked inheritance

Duchenne muscular dystrophy, haemophilia and **Fragile X syndrome** are all examples of genetic conditions with an X-linked recessive inheritance pattern. The most common form of colour blindness is also inherited in this way, though it is usually described as a trait, rather than a disorder (see box on p.48). X-linked recessive disorders usually affect only the boys

Colour blindness

Red–green colour blindness is not blindness at all but just a difficulty in distinguishing shades between red and green. It affects about one in twelve men, but only one in two hundred women. If you are a man with colour blindness, then you may have nephews or grandsons who also have this trait. Your children may all have normal vision, though all your daughters will be carriers of the trait. Female carriers may have sons with colour blindness, but not daughters. Your eyes detect colour with special cells that contain chemical pigments, and people with red–green colour blindness have pigment genes that are missing, or not working properly. Genes on the X-chromosome normally make these chemical pigments. As females have two X-chromosomes, they usually have at least one set of working pigment genes. But males have only one X-chromosome (and a Y-chromosome, which does not have any pigment genes), so if they have a single altered pigment gene they will be colour blind.

in a family. This is because they are caused by faulty or missing genes found on the X-chromosome. Girls are born with two X-chromosomes, so the effects of a working gene on one can mask the effects of a faulty gene on the other. They will normally be unaffected carriers of the condition. Boys have only one X-chromosome, so if they inherit one with a faulty gene, they will be affected. Any son born to a female carrier therefore has a one-in-two chance of being affected by the disorder, while each of her daughters has a one-in-two chance of being a carrier herself.

There are also a very small number of genetic conditions that have an X-linked *dominant* inheritance pattern – for example, some types of hereditary hearing loss – in which the effects of the faulty X-chromosome gene are *not* masked by a working copy. In this case, affected women will have a one-in-two chance of passing the disorder on to each of her children. An affected man will pass the condition on to all his daughters, but not to his sons (since they inherit only his Y-chromosome). Finally, some disorders show a Y-linked inheritance pattern, as they involved genes located on the Y-chromosome. Only men are affected by such disorders, and all the sons of an affected man are themselves affected. There aren't many genes on the Y-chromosome, so there are very few disorders that show this type of inheritance – although such mutations do explain some cases of male infertility.

Other types of inheritance

There are several other types of inheritance that do not fit any of the patterns described above. They are sometimes described as "non-Mendelian" to distinguish them from disorders and traits that have a straightforward dominant or recessive inheritance pattern. Some are variations of simple inheritance patterns. A dominant disorder will sometimes not affect everyone who inherits the faulty gene – a phenomenon known as **incomplete penetrance**. Other inherited disorders may not affect everyone in the same way, with different family members showing varying symptoms. In some genetic conditions, there is a tendency for the symptoms to worsen in each subsequent generation of the family, a situation known as anticipation.

A few diseases are caused by mutations in the genes of **mitochondria**, the cell's "powerhouses" that are inherited solely through the mother (see box on p.33). Lastly, some genes are described as being **imprinted** – that is, whether they are active or not depends on which parent they are inherited from. Only a small number of mammal genes are imprinted, a process that involves either the maternal or paternal copy of a gene being switched off very early on in development, and remaining switched off throughout the animal's life. A few genetic conditions involve imprinted genes, in which a failure to inherit a working version of the active gene can cause health problems (since the other version, inherited from the other parent, is switched off). Imprinting problems could also be behind the high failure rate of animal cloning, a topic covered in chapter ten.

Some of these ideas about genes depend on knowing exactly what these minute elements in the cell actually are; some do not. But the modern view of genes is firmly focused on the level of **molecules** – lengths of **nucleic acid**, and how the information they contain is organized. In the next chapter, we shall look at how genes are approached in this way in the lab.

Genes in the lab

So far we've laid out some of what we know today about genes, but we've glossed over exactly *how* scientists study tiny changes in the structure of a substance which, though comparatively big by molecular standards, is still too small to visualize without a very powerful microscope. But before we go on to some of the implications of genetic knowledge, for ourselves and our future descendants, it is good to know a bit more about techniques central to studying and manipulating DNA. Our intention here is to shed light on the possibilities and limitations of these often controversial technologies, to help separate fact from fiction before we consider both.

Studying DNA

Getting hold of the stuff

To study DNA, scientists first need to separate it from everything else that makes up a cell – mainly protein and fats. The methods for achieving this vary, depending on whether you are after **bacterial**, **viral**, **plant** or **animal DNA**, but generally involve the same chemicals. Take human DNA, for example. Since you have a complete copy of your genes in most of your cells, anyone who wants to study them can use a tissue sample from almost any part of your body. In practice, as for most medical tests, if doctors need your DNA they go for the easiest option: a few millilitres of blood. To be precise, they need your **white blood cells**, since red blood cells have no DNA in them.

The first step in a human DNA "blood prep" is to spin the sample at high speed, to separate the blood cells from the blood fluid. The cells are then mixed with a solution that causes them to burst, releasing their contents. The sample is spun again to separate out the **nuclei** – the structures at the centre of each cell that contain the vast majority of its genetic information. Next, the DNA is released from the nucleus using a mild detergent and a protein-destroying **enzyme** (getting fats and protein out of your DNA is little different from getting them out of your dirty washing). After some form of "cleaning up" to remove the resulting debris, alcohol is added – since DNA is not soluble in alcohol this step causes it to appear as a sticky blob in the mixture. You may think this all sounds quite straightforward, which it is – schoolchildren can easily get DNA out of a variety of things, including onions, peas and fish eggs. (See www. scienceinschool.org/2006/issue1/discoveringdna for a great example.)

In the last twenty years, the ability to amplify tiny amounts of DNA (using a technique called **PCR**) has meant that a single hair or a few cells scraped from the inside of your cheek will provide enough genetic material for a forensic, paternity, ancestry or health test, as we discuss in chapter 11. But for analyzing DNA in detail, larger, purer samples produced by methods similar to that described above are still the norm, although many labs now use commercial kits and machines to speed up the process.

Handling DNA

Once scientists have a solution of pure, clean DNA in a test tube (usually a small plastic tube with a lid), the next step is visualizing the particular

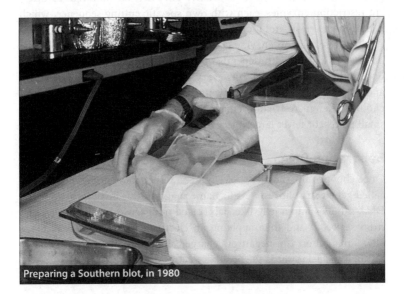

Preparing a Southern blot, in 1980

piece they are interested in. Before it was possible to amplify specific regions of DNA using PCR, this would involve chopping up and separating the resulting DNA fragments according to size, before transferring them to a nylon membrane for further analysis. This technique – called **Southern blotting**, after its inventor, Oxford geneticist **Ed Southern** – was the workhorse of genetics (particularly **human gene hunting** studies) from the 1970s right through to the early 1990s. It is not used so much today, but the principles involved form the basis of many other gene manipulation methods. (We'll explain the actual mechanics of the process shortly.) In one of biologists' little jokes, a similar technique for separating RNA molecules is called **Northern blotting**, while **Western blotting** is used to analyze proteins (as yet there's no such thing as an Eastern blot).

Chopping up DNA

Chopping up DNA requires special proteins called **restriction enzymes**. Originally discovered in bacteria in the 1970s, these "molecular scissors" cut DNA at precise points in its sequence – a handy defence against the viruses that regularly attack bacteria. They are named after the bacteria in which they were first identified. **EcoR1**, for example, comes from *Escherichia coli*, better known as one of the main microbes living in your gut. EcoR1 will snip DNA whenever it comes across the sequence of **basepairs** ("letters" of the DNA code) GAATTC. There are hundreds of different restriction enzymes, each with own specific cutting sequence, which makes them very useful tools for mapping and manipulating DNA.

When a restriction enzyme cuts up DNA, the number and size of the pieces you end up with depends on how often its cutting site turns up. For example, while EcoR1 cuts every time it finds the letters GAATTC, another (called Alu1) cuts every time it finds AGCT. So scientists can choose different restriction enzymes, to cut up DNA into either lots of small pieces, or fewer larger pieces.

Separating DNA

Once chopped up, the pieces of DNA can be separated using a technique called **gel electrophoresis**. As its name suggests, this involves loading the DNA samples into a jelly-like slab (made of **agarose**, which is extracted from seaweed) and then applying an electrical current across it. Since DNA in solution is a negatively charged molecule, the fragments move through the gel towards the positive electrode.

An agarose gel can be used to separate different sized fragments of DNA

The gel acts as a kind of molecular sieve, through which smaller fragments can pass more easily than larger ones, so the DNA is separated by size. Depending on the particular type of electrophoresis used, an agarose gel can separate fragments of DNA from about fifty base-pairs up to several million base-pairs (**megabases**). Since DNA is colourless, the gel is stained with some kind of dye, usually a chemical called **ethidium bromide**, which makes DNA fluoresce pink under UV light.

When viral DNA is cut up and separated in this way, you get a set of discrete "bands", since the complete genetic information of a typical virus (its genome) is only made up of, say, 50,000 DNA base-pairs (50 kilobases, kb). However, since the human genome is around three billion base-pairs in size, cutting up and separating human DNA results in a "smear" made up of around 1.5 million DNA pieces of different lengths.

Picking out a piece of DNA

To study the separated DNA in more detail, it can be transferred from the gel to a piece of **nylon membrane**. Before this, the gel is treated to make the DNA "single-stranded" – that is, the two strands of DNA that make up the double helix are "unzipped" (see chapter 1). Then comes the aforementioned **Southern blot**. The DNA on the gel is transferred to a piece of nylon membrane, usually by the endearingly low-tech method of placing it under a pile of paper towels with a heavy weight on top and leaving it overnight. Then scientists can pick out a particular piece of DNA on the nylon membrane using a **probe**, a small piece of DNA – also unzipped to make it single-stranded. The probe, tagged with either a chemical dye or radioactivity so it can be detected, is washed over the membrane for several hours.

The unzipped DNA probe will stick to any matching pieces of DNA on the membrane by a process called **hybridization**, which relies on the unique ability of DNA to stick to a "complementary copy" of itself: the chemical letter A will always pair up with a T, and a C with a G.

So, for example, GATTACA will stick to the complementary CTAATGT. This is the key to many techniques that aim to pinpoint one specific gene or piece of DNA – which would otherwise be like trying to find a needle in a very large haystack. If the probe is radioactive, then the probe-treated membrane needs to be exposed to a piece of X-ray film, which, when developed, reveals the location of the DNA of interest.

The original **DNA fingerprinting technique** relied on Southern blotting, followed up by hybridization with probes that detect sections of variable DNA – producing a unique "barcode" for an individual (for a detailed account of DNA fingerprinting see chapter 12). Some gene mutations that cause disease are also detected this way – for example, the expanded piece of DNA code responsible for the common genetic condition **Fragile X syndrome**. But for most forensic and medical applications, Southern blotting and subsequent hybridizations have now been replaced by PCR-based methods and DNA microarrays.

DNA microarrays

While Southern blots allow the detection of one particular piece of DNA at a time, **microarrays**, also known as DNA "chips" allow many different pieces to be examined simultaneously. Microarrays are in effect the reverse of Southern blots, in that many individual "probes" are immobilized on the chip, to which a sample of labelled DNA (or RNA) is then applied. To make a DNA chip, scientists cram ordered fragments of DNA (for example, gene fragments known to be altered in breast cancer) on to a piece of glass, plastic or silicon, roughly the size of a thumbprint. They can then investigate which versions of the genes are present in a sample of DNA from a patient simply by applying the sample to the chip and seeing if they react with the reference fragments. The test sample of DNA is first unzipped by heating, then labelled with a fluorescent dye, and washed over the chip. As with DNA probes sticking to DNA on nylon membranes, chips rely on the natural ability of DNA to find a complementary copy of itself. So any gene in the sample that matches a fragment on the chip will stick to it, and show up as a brightly coloured dot.

As well as their use in screening for many different mutations at once, other sorts of microarrays can be used to pick up large-scale deletions and duplications in a sample of patient DNA, though they have not yet replaced more traditional, cheaper methods of doing this, which rely on looking at whole chromosomes under the microscope. Another area in which microarrays are widely used is in the study of **gene expression** – that is, investigating where, when and to what extent different genes are switched on. Microarrays that contain all of the 6000 or so genes of the **yeast genome** have been available since 1997, and it is now possible to get two microarrays representing the entire human genome, around 25,000 genes. Using these, it is possible to look at gene activity in any sample of tissue – for example, RNA from a healthy liver can be compared with a sample from a cancerous liver, to find out which genes are involved in the disease process.

Reading the DNA code

So much for looking at pieces of DNA and the differences between them – peering ever closer, the next step is to read the actual genetic code, to determine the order of the base-pairs (ie As, Cs, Gs and Ts) in a particular section of DNA. The simplest method of **DNA sequencing**, as this process is called, was invented in the UK by double-Nobel-prizewinning scientist **Fred Sanger**. Such was his contribution to modern genetics research that the UK's Human Genome Project centre in Hinxton, Cambridge is named after him (www.sanger.ac.uk). The Sanger method of DNA sequencing relies on the natural ability of DNA to copy itself, using a **builder enzyme** called **DNA polymerase**. The polymerase makes many *partial* copies of a piece of single-stranded DNA, and copying is halted at either an A, C, G or T base. The technique works by repeating this process in four separate tubes – one with copies stopped at all the As, one at the Cs, one at the Gs and one at the Ts.

Once scientists have generated many partial copies of a piece of DNA – a mixture of different sizes because they have been stopped at different points in the sequence – they need to separate them out using a form of gel electrophoresis (see p.52). A very thin gel made out of a substance called **polyacrylamide** is used, as it separates smaller DNA fragments (between

Walter Gilbert (1932–) and Fred Sanger (1918–)

American Walter Gilbert and Englishman Fred Sanger offer a study in contrasting scientific careers. Both won a share of a Nobel prize for DNA sequencing techniques in 1980. For Sanger, it was an almost unprecedented second Nobel triumph, the culmination of a life in the lab trying to unravel nature's most complex molecules. For Gilbert, the prize, and the sequencing work, were an important episode in a life deeply involved in public affairs, business, and helping to shape the future of his science.

Fred Sanger was the son of a country doctor in Gloucestershire, but was drawn to science more than medicine because he thought it would provide a better focus on a single set of problems. After study at Cambridge, he stayed on to do a PhD after the beginning of World War II (he registered as a conscientious objector), working first on metabolic biochemistry, then on the small protein **insulin**. This eventually led to his first great feat of chemistry, breaking insulin up and using the fragments to piece together the sequence of **amino acids**. By the time he was 40, he had been awarded his first Nobel prize. Some Nobel winners happily assume a public role. Sanger did the opposite. Unhappy with teaching or administration, he used his status to reinforce his position as a man who stayed in the lab and did his research.

Still in Cambridge, he joined the stellar band that founded the Medical Research

Twice-Nobel-winner Fred Sanger, examines a model of a DNA molecule, c.1980

10 and 1000 base-pairs) than agarose gels do. Again, the shorter the piece, the faster it travels, so the pieces in the gel are arranged according to their size, revealing the order of the As, Cs, Gs and Ts in the DNA strand. This method originally involved tagging the fragments with a radioactive label.

Council's Laboratory of Molecular Biology in 1962. "In this atmosphere", he recalled, "I soon became interested in nucleic acids." And so he and his colleagues figured out how to sequence them, first with chemical methods for sequencing smaller pieces of DNA.

The first organism to be sequenced in 1977, a minute bacterial virus, had just over 5000 nucleotides. But it proved the method, and Sanger's technique became the basis of the huge banks of automated sequencers which were the mainstay of the Human Genome Project. His group also developed the "shotgun" method for larger genomes, which remains the other mainstay of mass sequencing. He retired a few years after the second Nobel award, but not before his group had sequenced the first human genome, albeit the tiny genome of **mitochondria**, in 1981.

Walter Gilbert trained in physics and was working as a theorist at Harvard in the late 1950s. However, he had met James Watson some years earlier and found himself drawn into the struggle to understand the genetic code. A summer working on messenger RNA led to a permanent shift in his research and he was soon working at the leading edge of molecular biology. He soon made a name with work on the isolation of the small protein – the **"repressor"** – which controls synthesis of an enzyme on the bacterium *E.coli*. It confirmed Jacques Monod's and Francois Jacob's theory of genetic regulation, and was the first of the vast array of gene control elements now known.

In the 1970s he worked on DNA sequencing, with results that led to his shared Nobel in 1980, and also became deeply involved in the new biotechnology. Exchanging his lab coat and bad haircut for a suit and tie, he was a co-founder of the pioneering biotech start-up Biogen, which used recombinant DNA methods to make **human insulin** (see chapter 9). He eventually quit Biogen and returned to teaching and research at Harvard, but became involved with several other companies, not least the short-lived Genome Corporation, which planned to sequence human DNA in 1988, and Myriad Genetics.

Gilbert was one of the strongest advocates of the later **Human Genome Project**, known for brandishing a CD in his talks and telling audiences they would soon have their personal genome on a disc. When the project got properly under way, he pressed for development of new computer techniques along with the accumulation of sequences, and wrote many articles outlining his vision of a biology which was rooted as much in databanks as laboratories. This stemmed partly from his earlier work on **introns** and **exons**, and his interest in evolution of genes and genomes. He remains professor emeritus at Harvard.

After running the gel it would have to be 'dried down' and exposed to X-ray film before the DNA sequence could be read, one base at a time – up to a maximum of 1000 base-pairs per gel. Chances are you will have seen one of these sequencing **autorads**, as they are called, as they have been widely adopted by newspaper articles and TV programmes to illustrate anything on human genetics.

Automated DNA sequencing

The decision to sequence the entire *three billion* base-pairs of DNA that make up the human genome understandably triggered a search for faster sequencing methods. Most DNA sequencing these days is done by tagging the DNA fragments with fluorescent dyes, rather than by radioactivity, using four different colours – one for each DNA base. This system allows the sequence to be read automatically, saving a lot of time and reducing the potential for human error. The fastest machines pull the different sized fragments through a very fine tube, or **capillary**, rather than using a gel. A laser beam then detects the moving, fluorescing DNA fragments as they pass, and produces a handy printout of the results.

The "reference" human genome sequence, the first draft of which was published in 2002, cost an estimated $3 billion, with mammalian genome projects today costing between $10 million and $50 million. The race is now on to develop next-generation DNA sequencing technologies, capable of whizzing through entire genomes both quickly and cheaply. In 2004, the **US National Institutes of Health** called for proposals for new methods capable of delivering an entire genome sequence for just $1000 – something it believes is achievable by 2014. One of the most promising new techniques is called **pyrosequencing**, which generates flashes of light every time a DNA base is analyzed.

Amplifying DNA: a revolution in genetics

It's no exaggeration to say that the polymerase chain reaction, or **PCR**, has revolutionized the science of genetics and its applications. PCR provides a way of rapidly amplifying any piece of DNA, providing most of its sequence is known, and it's not too big – a few hundred base-pairs works best. Maverick US scientist **Kary Mullis** (see box on p.60) invented this powerful new technique in 1983, and it came into widespread use by the end of the decade

Using PCR, scientists can obtain plenty of DNA from just a few cells in an hour or so, and for this reason it has transformed forensic and medical genetic testing. Because it only requires a tiny amount of starting DNA, it has also allowed a whole new area of science to blossom, in the form of **molecular paleobiology** – the study of DNA from ancient specimens of bones and teeth. It was this aspect of PCR that provided the inspiration for the book and subsequent film *Jurassic Park*, in which whole dinosaurs are recreated using DNA from dinosaur blood preserved in the guts of prehistoric insects trapped in amber.

In reality, even DNA, an unusually stable molecule, cannot survive for the millions of years necessary to make *Jurassic Park* a reality. Short fragments of DNA may last for 50,000 to 100,000 years before water, air and microbes finally destroy them. There were a few high profile reports in the early 1990s from scientists who claimed they had isolated DNA from much older specimens, but later analysis revealed that these studies had largely been picking up tiny amounts of contaminating modern DNA. But amazingly, PCR *has* been used to amplify short segments of DNA from woolly mammoths, ancient mummies and Neanderthals, amongst other things.

Like DNA sequencing, PCR relies upon the natural ability of DNA to copy itself, using a special type of **DNA polymerase**. During PCR, the enzyme copies a section of DNA in the sample millions of times using the bases A, C, G and T as the raw materials, and short strands of DNA called **primers** to start the process. So one molecule of DNA quickly generates two more, which in turn provide template strands for four more, and so on, for 20–25 "cycles". You can watch an animation of this process, along with some of the other techniques described in this chapter, at the **DNA Learning Center** website – the educational wing of the US Cold Spring Harbor Laboratory: www.dnalc.org/ddnalc/resources/pcr.html

The key to the runaway success of PCR is the use of a super heat-resistant type of the polymerase enzyme. To carry out PCR, scientists use a machine that heats the DNA to 95°C. At this temperature, the DNA unzips into two separate strands, which the polymerase uses to make more DNA copies. Most enzymes (being proteins) are destroyed at this temperature, with the exception of those found in bacteria that live in hot water springs. By using a polymerase from one such microbe, *Thermus aquaticus*, PCR went from being a laborious manual procedure, in which fresh enzyme had to be added every cycle, to one that could be fully automated. The development process, in which the relative contributions of Mullis and his former colleagues at the biotech company Cetus remain controversial, is chronicled in Paul Rabinow's excellent 1998 book *Making PCR*.

Once generated, the products of a PCR reaction can be treated in the same way as any other piece of DNA – cut up with a restriction enzyme, run out on a gel, used to make a DNA probe, checked against a microar-

Kary Mullis (1942–)

Colourful is the adjective usually used for Kary Mullis, a much-married chemist who has turned his skills to synthesizing psychoactive drugs as well as dealing with DNA. After a series of university appointments in the US, his career was relatively unremarkable until he signed up with one of the new Californian biotechnology companies, Cetus, in 1979.

At Cetus he set to work pondering ways of improving the efficiency of methods for making short stretches of DNA (**oligonucleotides**), and of using them to detect key sequences in DNA samples. Working with the enzymes which make DNA in the living cell, he saw a way to put together a procedure for amplifying small samples of DNA, making many copies of the same sequence.

The basic idea, which involves repeated cycles of heating and cooling to separate DNA strands, came to him, he relates, while driving a southern California highway in 1983. As it revolves around use of the key enzyme **DNA polymerase**, it was dubbed the **Polymerase Chain Reaction** (PCR).

Automated PCR machines are now basic kit in any DNA lab, but it took years to develop the idea into a reliable technology. A crucial improvement came when Mullis realized the method could use an unusual DNA polymerase from

a bacterium whch lives in hot springs, and which tolerates high temperatures. Mullis and Cetus later disputed his share of the credit, as well as the patent rights to PCR, but his claim to the idea was endorsed by his rapid elevation to the Nobel pantheon in 1993, with the prize for chemistry.

Since his departure from Cetus, he has worked for other companies, been a consultant, lecturer, expert witness in DNA cases, and a critic of the evidence that HIV is the main cause of AIDS. His 1998 autobiography, *Dancing Naked In The Mindfield*, presents a rather self-regarding maverick, intent on going his own way, who sounds more amusing as a friend than a colleague. But he certainly shows that, with luck, you can be a successful scientist without toeing anybody's corporate line.

ray, sequenced or even cloned. There are several clever PCR-based methods that allow the detection of single base-pair mutations in any piece of amplified DNA. For this reason it has largely replaced **Southern blotting** as the main technique used in medical genetic laboratories. It is also a central tool in other areas of medicine, for example identifying microbes that cause disease, and diagnosing cancers.

Cloning DNA

There are two types of cloning. There is the sort that aims to generate an identical genetic copy of a whole organism, of whom **Dolly the sheep** is the most famous example (and who graces the cover of this book). We'll cover all that in chapter 11. First, we'll look at DNA cloning, the technique that is the basis of "genetic modification" or **GM technology**. DNA cloning is basically a "cut and paste" technique for getting a particular piece of DNA out of one organism's genome, and into that of another organism. In the days before PCR, cloning DNA in fast-growing microbes was the only way of generating many copies of a particular piece of DNA for further study. The sequencing of the human genome, for example, relied on the painstaking analysis of hundreds of thousands of different bacterial and yeast clones, isolated from clone "libraries".

To clone a piece of DNA, it must first be excised from the surrounding DNA. This always used to be done using a **restriction enzyme**, but nowadays can also be done using PCR. The fragment of DNA is then inserted into a **vector** – a "carrier" that will enable the cloned DNA to be accurately and efficiently copied once it's inside the host organism. The first cloning vectors used were **plasmids** – tiny circular pieces of DNA found naturally in bacteria, where they carry useful genes such as those that confer resistance to antibiotics. The target and vector DNA (cut up with the same restriction enzyme) are mixed up with another enzyme called **DNA ligase**, which sticks the two together. This process of creating novel DNA sequences is known variously as **recombinant DNA technology**, genetic engineering or gene splicing. The new, genetically altered plasmid can then be put back into the host strain of bacteria, where it is copied along with the rest of the host DNA every time the cell it's in divides.

To make a whole library of clones, the entire genome of an organism can be chopped up and added to the vector DNA, resulting in lots of different clones that can be individually characterized. As you can imagine, for the **Human Genome Project**, this meant a lot of very repetitive work, as the DNA in each bacteria and yeast clone had to be completely sequenced and compared to the others (see chapter 4 for the full story of this massive undertaking). Another very useful sort of clone library is one that just contains copies of the genes that are switched on in a particular tissue, rather than all the DNA of an organism. So-called "complementary", or **cDNA** clones are made by copying all the RNA present in a tissue sample – the intermediary molecules between genes and the proteins they code for. Such libraries have proved invaluable in the identification of genes involved in human diseases – the search for a gene involved in a muscle disorder can be considerably narrowed down by looking in a cDNA library made from muscle tissue, for example.

Making GM plants and animals

Having a piece of DNA in a microbe means that you can make lots of it, simply by growing the **bacteria** or **yeast strain** that contains it. As well as allowing scientists to study individual genes in detail, this technology has meant they can get bacteria to mass-produce proteins that they would not normally make, turning them into mini **bioreactors**. For example, the enzymes added to washing powders are often churned out by genetically engineered bacteria, as is **chymosin**, a vegetarian alternative to rennet (a protein found in calves' stomachs) used in the manufacture of cheese.

In medicine, there are several proteins now obtained from bacteria that previously came from cadavers or animal tissues. For example, all the insulin used to treat people with **type 1 diabetes** used to come from the pancreases of cows and pigs. Many patients now receive human insulin produced by bacteria – one of the first commercial success stories of recombinant DNA technology, as we describe in chapter 10. Other therapeutic human proteins, such as human growth hormone and **interferon** are also now made by bacteria.

Moving beyond bacteria, genetic engineering provides a way in which genes can be "shuttled" between different animal and plant species – a process termed **genetic modification** (GM). Since nearly all living creatures share the same genetic code, in theory it should be possible to take a gene from one species and, provided it is accompanied by the appropriate control DNA sequences, it will produce exactly the same protein in its new location. GM plants can be made by introducing the required gene – say, one that confers drought resistance – using a soil bacterium carrying a plasmid. The naturally occurring **Ti plasmid** is often used to create GM plants, because it has a natural tendency to invade plant cells and insert its own DNA into the plant DNA. However, plant biologists have had to come up with other ways of getting novel genes into plants, as Ti plasmids don't work for crops such as rice, corn and sugar cane. Other methods include using tiny metal spheres coated with the gene of interest, or subjecting plant cells in a laboratory dish to small electric shocks, allowing the gene to enter.

Scientists can breed genetically modified animals too, though the techniques involved and reasons for doing so are different from those for GM plants. To create a GM, or **transgenic animal**, the new gene can be directly injected into the nuclei of newly fertilized eggs in the laboratory – a tricky technique called **microinjection**. Any that survive this process and manage to start developing into an embryo are then returned to the womb of a host female. Many thousands of different transgenic mouse strains have been created in this way, mainly to shed light on human diseases or normal embryo development. While some have human genes inserted, others have crucial genes knocked out, in order to find out what their usual role is in the body. When you see figures recording recent increases in the use of animals in laboratory experiments, the large majority of them are specially created transgenic mice.

Another way of creating a transgenic animal is by using cloning technology, in which the gene is first added to the genetic material of an ordinary cell, rather than an egg cell. The entire cell nucleus is then transferred to an egg cell that has had its own genetic material removed, which is then given an electric shock to make it start developing (the Dolly method). **Polly**, the first GM cloned sheep was created this way – although she is much less well known than her feted predecessor, her birth was arguably more scientifically significant. A third way to make genetically altered animals is to insert genes into **embryonic stem** (ES) cells growing in the lab – cells from a very early stage of development that still have the potential to grow into any type of body tissue. The genetically altered ES cells are then fused with an early embryo and allowed to develop. This process actually results in a **chimeric**, rather than a transgenic animal, since its cells will be a mixture of ones that contain the new gene and ones that don't (the word chimera comes from the mixed-up creature of Greek mythology, which had the body of a goat, the head of a lion and the tail of a serpent).

Apart from their use in basic research, there has been great interest in recent years in creating larger transgenic animals for **pharming**. This is when human genes are inserted into animals in the hope of harvesting large quantities of the protein they make – a scaled-up version of bacterial bioreactors. For example, a new drug based on a human protein extracted from the milk of genetically altered goats was approved for clinical use by European regulators in 2006. It contains the **antithrombin** protein, which is used for treating patients affected by a rare inherited blood clotting disorder.

Finally, the ability to get human genes working in other animals might one day address the shortage of human organs available for transplant operations. One suggestion is to use pig hearts, since they are roughly the same size as human hearts. Scientists have bred transgenic pigs whose heart cells carry key human proteins on their surfaces, with the aim of fooling the human immune system to help prevent rejection. However, there are concerns about the safety of transplanting organs from animals to humans (**xenotransplantation**), especially the risk of infection by animal viruses. There are also worries that we do not know enough about what genes (and the proteins they make) do when they are transferred from one animal to another, so people are unlikely to be receiving pig organs any time soon.

The whole array of techniques we have described began in the lab, as aids to studying genes. But as we have seen they soon began to make their way into the wider world – both into other labs, where they found other uses within biology, and into applications which spawned a whole range of new technologies in medicine, agriculture, industry, and even crime-fighting. Looking at some of these in more detail takes up much of the rest of this book, beginning with some of the scientific spin-offs of the new genetics.

PART 2
GENETIC
LANDSCAPES

Decoding humans
(and other animals)

"Biology's Apollo landing", "the outstanding achievement of human history" and "the language in which God created life" – these are just a few of the comments that accompanied the unveiling of the rough draft of the **human genome** in June 2000. By April 2003, when the international scientific team published its "final" version (actually just another draft) the media attention had diminished slightly. However, scientists and governments alike continued to promise that genetic knowledge would transform medicine. So how did a stream of As, Cs, Gs and Ts cause such a stir, and has the Human Genome Project lived up to all the hype? In this chapter, we'll look at how geneticists worked in the pre-genome era; the reasons for reading the entire genetic code; how they tackled this Herculean task; and how they are now making sense of the resulting three billion base-pair DNA sequence.

Finding tiny needles in a big haystack: identifying human genes

Human gene identification started off slowly, with just a couple of hundred of them being pinpointed during the 1970s. Many of these early successes were genes found to be altered in people affected by diseases with a known faulty protein – for example, those that are responsible for inherited blood disorders such as **thalassaemia**. Scientists could work backwards from the order of amino acids that make up the protein to its corresponding DNA code. In fact, many human genes have been identified solely by their involvement – when missing or faulty – in some kind of disease or abnormality. This in turn means that many scientists and journalists alike now refer to "the cystic fibrosis gene", or "a gene for diabetes". But as geneticists Tom Strachan and Andrew Read have pointed out, labelling genes by what happens when they don't work properly is a little misleading – it's like calling your freezer "a machine for ruining frozen food". It also wrongly implies that only affected people have the gene, when in fact *everyone* has two copies of almost every human gene (barring a few located on the Y-chromosome), it's just that, for example, "the cystic fibrosis gene" is missing or altered in people with cystic fibrosis. This oversimplified view of what genes do is something we shall return to at the end of this chapter.

Rapid advances in technology during the 1990s meant that it became possible to identify mutated genes responsible for human disease even if nothing was already known about the protein they coded for. This triggered something of a genetic gold rush, in which the prizes were publications in top scientific journals. By 1998 over 6000 human genes had been identified – including those responsible for the relatively common genetic conditions **cystic fibrosis** and **Duchenne muscular dystrophy**. These projects typically took several years, since at that time the vast majority of the human genome was uncharted territory – the search for the **huntingtin gene** (the gene which, when mutated, causes Huntington's disease), for instance, lasted a decade. It's worth considering just what an arduous process 20th-century gene hunting was before we look at how things have changed in the last few years.

Hunting for a gene

Trying to track down a particular disease gene is like trying to find one particular building in a big capital city with just a photograph to help you. The entire human genetic code is made up of about three billion base-pairs of DNA – that's 3,000,000,000. Humans have about 25,000 different genes, each made up of between a few hundred to a few thousand base-pairs of DNA code. These genes are bundled up into 23 pairs of chromosomes, each of which carries thousands of different genes and is made up of millions of base-pairs of DNA code. If you had to look for a building in New York, it would help to find out which district it was in, then which street, and finally its street number. Likewise, gene hunters first have to find out which chromosome their gene is on, then which *part* of the chromosome it is located within and, finally, identify the right gene.

Sometimes, careful analysis of a patient's chromosomes can provide vital information – part of a particular chromosome may be missing, rearranged or even duplicated. But if this is not the case, then scientists must rely on the use of "reference" fragments of DNA: they study how a genetic condition is passed on from one generation to the next, and look for a piece of DNA that is inherited along with the condition. Tracking disease genes across affected families in this way is a procedure known as **linkage analysis**. During the search for the gene involved in **Huntington's disease**, for example, a DNA fragment from chromosome 4 gave researchers the first clue they needed. These reference fragments – so-called **polymorphic DNA markers** – come in several versions, so that the chromosomes carrying the normal gene can be distinguished from those carrying the faulty gene.

If the words "polymorphic DNA marker" seem a little opaque, you might like to consider the process in terms of one simple, variable inherited feature – hair colour, say. If, over several generations, affected individuals always have blonde hair, while unaffected individuals never, or only occasionally, have blonde hair, then it might reasonably be assumed that the gene involved in the disease is physically close to the blonde hair gene. If they are extremely close together on the same chromosome, then they may never be inherited separately. However, if they are located more than a few hundred thousand base-pairs away from each other, then they will sometimes be separated during the chromosome shuffling that goes on during egg and sperm production (a process called **recombination**: see p.29). By measuring how often the two are inherited separately, scientists can estimate the genetic distance that separates two traits – in this case, hair colour and the disease gene. Genetic distances are measured in **centimorgans** (cM), named after the US geneticist **Thomas Hunt Morgan**, who constructed the very first genetic maps (of fruit fly chromosomes) in the early part of the twentieth century. In humans, 1cM is roughly equivalent to 1,000,000 base-pairs of DNA – 1megabase (Mb) – but the relationship between physical and genetic distances isn't always straightforward. Some chromosome regions seem particularly prone to breaking, and DNA markers or traits located either side of these so-called "recombination hotspots" may seem to be further apart genetically than they are physically.

But pinning down the chromosome location is just the first step in a gene hunt – it's like identifying which district your building is in. The largest of the human chromosomes, **chromosome 1**, is made up of about 300 million base-pairs of DNA code. Even the smallest,

The race for the "breast cancer gene"

Breast cancer is appallingly common in women (about as common as prostate cancer in men), and it has long been known that it can run in families. But useful information about genes and disease was confined to relatively simple cases, like **Huntington's** or **cystic fibrosis**. Breast cancer was far from simple, and there was little reason to believe that the tangle of genetic and environmental factors involved could be unravelled.

That changed in 1990, when **Mary-Claire King**, then at the University of California at Berkeley, revealed results from fifteen years of work with 1000 women of Ashkenazi Jewish descent who had a particularly high incidence of breast cancer. She was able to show that there was a single gene involved when these women got cancer, and that it was somewhere on the bottom half of chromosome 17.

This was the starting pistol for a race to locate the gene more precisely, and then clone it. The contenders included King herself, Mark Skolnick of the University of Utah and the biotech company Myriad Genetics, and researchers at the Institute of Cancer Research in London. At stake was not just the medical interest of finding out what the gene might do, but the possible commercial bonanza from a successful test.

It turned a useful advance in medical genetics into a prolonged series of disputes about ownership. The team based at Myriad Genetics in Salt Lake City had

an advantageous local resource in the large families of Utah, and the Mormon Church's interest in genealogy. Some of the families they took DNA samples from had more than forty breast cancer sufferers. In August 1994, Myriad announced that they had succeeded in sequencing the gene, known as **BRCA1**, and in identifying the mutations that lead to an increased risk of developing breast cancer. Myriad filed for a patent on all tests based on BRCA1 that October – one that did not initially credit the US Government or its National Institutes of

Mary-Claire King

Health (NIH) as one of the "inventors", despite the fact that they had put $5 million into Mark Skolnick's research.

It is important not to exaggerate the significance of BRCA1. Only about five percent of breast cancers are thought to occur in women with one of the major risk genes, and BRCA1 mutations account for about half of those. And the choices for those who do carry the gene – regular mammograms or perhaps a double mastectomy before cancer occurs – are hard. Nevertheless, a BRCA1 mutation can increase the lifetime risk of contracting the disease to around 80 percent, so it is of course an extremely significant gene for the families involved.

Several arguments ensued. The first, and simplest, was with the NIH, which had supported early work on the gene and filed a separate patent claim. This was quickly resolved by the two agreeing to split the rights. Further protests came from some of the women who had given DNA samples to Myriad. They had volunteered to help locate the gene and ultimately move towards new treatments and cures for breast cancer, and many were angry that Myriad had patented the gene and any tests that used the sequence they discovered – thus making it illegal for other companies to offer diagnostic tests except under licence.

Further disputes followed, over whether Myriad should have rights to a particular test, a range of possible tests, or over the entire gene. For example, the company resisted attempts by the British National Health Service to use a simpler, and cheaper test that did not sequence the entire gene. Things became even more confusing when Myriad announced it claimed rights to a second gene implicated in breast cancer, BRCA2, the day *after* details of the gene were published by a British group based at the Institute of Cancer Research. It appeared that the company had filed their claim mere hours before the ICR's announcement. Myriad maintained that their sequence was more complete and a better basis for testing.

The story that unfolded involved other companies and laboratories, as well as patent offices in the US, Europe and the UK, and was lengthy and complex. Initially Myriad retained US and European patents to BRCA1 and 2, while the ICR has a UK patent on BRCA2. But a series of hearings by the European Patent Office led to the patent awards on BRCA2 and then some of its rights over BRCA1 being struck out. So although Myriad retains some protection in Europe, there remains a transatlantic divide over who can do what, for whom, and at what cost. Meanwhile, study of the genes' function has produced important findings for cancer biology, and more than 100,000 women have been tested for one of the two genes. You can read about Myriad's current testing programme at www. myriadtests.com/mrap.htm

In the UK, your doctor can buy the test direct from Myriad for £1800, and get the results in a few weeks. Alternatively, it can be obtained through the National Health Service, although the charity and pressure group Breakthrough Breast Cancer claimed in 2006 that some patients were often waiting years for their results.

chromosome 21, is about 50 million base-pairs long. A single gene is usually only only hundreds or thousands of base-pairs long – just a fraction of the length of the whole chromosome. Pre-Human Genome Project, gene hunters had to narrow down their search by constructing a genetic map of the chromosome, with known landmarks on it – in other words, more reference DNA markers. Next, they had to zoom in on the area thought to be harbouring the gene of interest, and get it into a form in which it could be studied further. To do this, they had to sift through "libraries" of **chromosome fragments**, each one propagated in a separate bacterial or yeast strain (see p.61 for more on this DNA cloning technique). Then, by matching up the reference DNA markers on overlapping chromosome fragments, they could put a physical map together – like putting together lots of overlapping photographs to make one big picture. After a few years of this painstaking work, many geneticists were calling for a more efficient approach to these types of studies.

By the last stage of the gene hunt, the search should be confined to a "candidate region", hopefully no more than a couple of million DNA base-pairs in size. However, there may be many genes in a small section of a chromosome, just as there are usually several buildings with a particular postcode. Certain other experiments can help narrow down the search at this point: which genes are switched on in the relevant tissue, for example, in muscle, if a gene involved in an inherited muscle disease is sought? Do any of the genes bear a resemblance to any of their animal counterparts, about which more may already be known? Once they have selected the best candidate, scientists must look at the DNA sequence that makes up each gene. If it really is the gene they are seeking, then they should find gene mutations in the DNA of affected people, and not in the DNA of any unaffected controls.

The Human Genome Project

It might seem obvious, in retrospect, that gene hunting studies would have been much quicker and cheaper if the whole human genome was methodically mapped and sequenced, saving scientists from having to go after one gene at a time. But getting this massive undertaking off the ground took several years of planning, organization, and fund-raising. Two separate problems first prompted the idea of sequencing the entire human genetic code. The first was to understand the genetic changes involved in disease, particularly those that cause cancer, which occur at the level of the individual cell. As biologist **Renato Dulbecco** said in 1986: "We have two options: either to try and discover the genes involved in malignancy by a piecemeal approach, or to sequence the whole genome of a selected animal species." The second was a push for new technologies to look for possible inherited DNA damage caused by radiation, particularly the atomic bombs dropped on Nagasaki and Hiroshima at the end of World War II. The combined medical and political momentum ensured that the international **Human Genome Project** (HGP), which eventually began in 1990, secured funding from both the US National Institutes of Health (NIH) and the Department of Energy (DoE). The project's price tag – an estimated $3 billion – horrified its critics, who said it would drain resources from other biological research. Other dissenters questioned the usefulness of sequencing the entire genome, when only a small proportion of it was thought to encode genes. Why not go after the interesting stuff, and forget about the non-coding DNA? In fact, much of this so-called "junk" DNA is now thought to play an important role in regulating gene activity, although that wasn't known at the HGP's outset. Despite the objections, the project finally got underway – initially under the watchful eye of DNA pioneer **James**

Francis Collins

How did they do it?

The human genome is very big. If you had to read the entire human DNA sequence out loud then, even at the brisk rate of ten bases per second, you'd still be busy for nine-and-a-half years. So the HGP scientists had to break the genome down into a series of manageable tasks, using scaled-up versions of the techniques used to look for single genes outlined earlier. The human genome is packaged up as 24 different chromosomes, (numbered 1–22, plus the X and Y-chromosomes). To help them read the entire sequence, scientists had to take each one and make 24 individual **chromosome maps**. In fact, they had to create two different sorts of maps – genetic ones and physical ones. Some of the "landmarks" on these maps are genes, but many others are just short sections of variable DNA – both the **restriction fragment length polymorphisms** (RFLPs) used at first, and, **minisatellite markers** discovered later, which are much simpler to study. Both these types of DNA marker contribute to the natural genetic variation that exists between individuals, which is exploited in forensic and paternity tests. For genetic mapping purposes, they allowed scientists to measure the genetic distances (in centimorgans, see p.69) that separate them. Making these genetic types of chromosome map involved tracking the inheritance of the genes and DNA markers in many families to work out their chromosomal location and order. This work encompassed and subsumed much of the work done for individual gene hunts, carried out both before and during the HGP.

The genetic maps of each chromosome formed a basis for the creation of physical maps, which detailed the distance in DNA base-pairs between each landmark. This was done using a variety of approaches, and again incorporated much of the data produced by smaller labs in their intensive search for disease genes. Making more detailed physical maps involved breaking up all the chromosomes into fragments of DNA, and then cloning them in bacteria or yeast hosts, so that each fragment could be studied in depth. All the individual clones were then assembled into "**contigs**" by identifying the overlaps between them: for example, two clones may share a particular gene. As well as the genes and DNA markers already present on the genetic maps, the physical maps relied on lots of randomly generated bits of DNA code called **sequence tag sites** (STSs), which acted as

Watson, who became head of NIH's National Center for Human Genome Research in 1988.

Watson was replaced by Francis Collins in April 1993 and the name of the Center was changed to the **National Human Genome Research Institute** (NHGRI) in 1997. Its aims were to construct detailed genetic and physical maps of all the human chromosomes, before going on to read the entire DNA sequence of each. The HGP funding also helped pay for the development of new, faster technologies, and the sequencing of five other genomes – that of a strain of bacteria, a type of yeast, the fruit fly, roundworm and mouse. Finally, a portion of the HGP's budget was given over to considering the Ethical, Legal and Social Implications (ELSI) extra reference points. Many of these were *expressed* sequence tag sites (ESTs), which represent fragments of likely genes. To complete the project, scientists used automatic DNA sequencing machines to read the order of the DNA base-pairs in each clone, and with the aid of powerful computers and the maps, they reconstructed the DNA sequence of each complete chromosome. They repeated this process several times, to iron out any errors and try to fill the gaps resulting from troublesome sections of DNA. Despite these efforts the "finished" human genome sequence unveiled in 2003 still had 341 gaps.

Celera Genomics' **whole genome shotgun technique** was somewhat more straightforward. It involved breaking the DNA that makes up a genome into fragments of various sizes, ranging from 2000 to 150,000 base-pairs in length. An automatic DNA sequencing machine then read a few hundred base-pairs from both ends of each fragment, and computers were set to work looking for overlapping DNA sequences. Before it was applied to the human genome, this method was used to sequence several **pathogen genomes**, the first of which was *Haemophilus influenzae* in 1995, a bacterium responsible for a variety of infections.

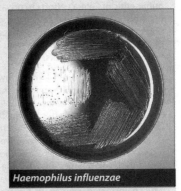

Haemophilus influenzae

However, despite its proven success in tackling smaller genomes, critics of shotgun sequencing feared it would not be able to cope with the vast stretches of repeated DNA present in the genomes of more complex organisms, because it would result in spurious overlaps. In the end, many scientists feel that the Celera effort relied heavily on the data produced by the HGP, which was updated every 24 hours and posted on the Internet for anyone to use.

raised by human genetic knowledge – for example, the potential misuse of such information by employers and insurers, and equality of access to genetic tests. There's a wealth of further information on the history, technology and implications of the HGP available on this DoE-maintained website: www.ornl.gov/sci/techresources/Human_Genome/home.shtml

Although much of the focus (and publicity) has been on the efforts of US scientists, the decoding of the human genome was very much an international effort, coordinated by the **Human Genome Organization** (HUGO) set up in 1988. Researchers based at the UK's Wellcome Trust Sanger Institute in Hinxton, Cambridge, contributed around one third of the data used to produce the draft human genome announced in

John Sulston (1942–)

While the public and private efforts to map the human genome were competing intensely, and arguments about the ownership of the information raged, defending the public interest largely fell to the British. Publicly supported scientists in the US felt as strongly about the right of free access to the information, but their government did not want to be seen to criticize free enterprise in the shape of Craig Venter's Celera Genomics. In the UK, conveniently, much of the money behind the national genome effort came from the independent **Wellcome Trust**. And their research leader, John Sulston, became the leading public critic of Celera's efforts to privatize human DNA. For the press and media, the personal aspects of the race became irresistible. Venter (see p.79) was brash, flash, American, and out to get rich. Sulston was British, white-bearded, pacific, unmaterialistic, and idealistic. And he wanted to keep our DNA out of the hands of the capitalists.

His earlier career did not foreshadow his media stardom during the genome wars. Born in 1942 in Cambridge, England – the son of a priest and a schoolteacher – he basically spent his whole life in the lab in the same city, save for a brief stint in California after his PhD. He came back from there in 1969 to join the Medical Research Council's Laboratory of Molecular Biology, and work with Sydney Brenner on the nematode worm *C. elegans*. Brenner had a far-reaching vision of the tiny worm as a key model organism for studying the action of genes in development. Sulston's role was to follow the division of the worm's cells. The fully grown C. Elegans always has 959 cells. After a dozen extraordinarily painstaking years' work, Sulston and his colleagues had a complete map of how divisions of the fertilized egg give rise to these cells. Along the way, they made important discoveries about the programmed cell deaths – 131 of them in this case – now known to be a vital part of the growth of all complex organisms.

After this feat, he switched to mapping and then sequencing the worm's DNA, still as a backroom boy. But as the tempo of sequencing increased, he gradually took on more management, and finally assumed the directorship of the new state-of-the-art gene sequencing complex at Hinxton Hall in Cambridge, funded by Wellcome and named after **Fred Sanger**. There followed the public dispute with Venter, and the eventual joint announcement of the "completion" of the project, following which Sulston quickly stood down as director, wishing to return to research. Sulston was awarded the Nobel prize for Physiology or Medicine in 2002, jointly with Sydney Brenner and his US colleague Bob Horvitz. Officially, their award was for the work on development of the worm, but for the press it was an endorsement of the genome project's public champion. In the wake of the award, he continues to comment on the uses of genetic information, and is vocal on the necessity of ensuring that the benefits from its use are shared equitably across the globe.

His own account of the genome saga, written with science writer Georgina Ferry, is called *The Common Thread* (2002).

2000. Before tackling the human genome, Sanger Institute director **John Sulston** cut his teeth, so to speak, on the roundworm genome (a mere 100 million DNA base-pairs long). As well as the US and UK, scientists based in twenty other countries took part in the HGP, most notably in Japan, China, France and Germany.

A lot of the media coverage of the HGP focused on the rivalry between the public sector effort and that of US biotech firm **Celera Genomics**, who were comparative latecomers to the research. This company, headed by maverick scientist **Craig Venter**, began its own human sequencing project several years after the HGP started. They took a different approach to generating the data, called "whole genome shotgun sequencing" which was much faster, although less reliable than the methodical approach being taken by the HGP (see box on p.74).

Celera's intention was to sell its annotated version of the human genome to researchers, a move that angered many public sector scientists who had always made their data freely available via the Internet. Venter had already incurred the wrath of **James Watson** during his time as an NIH researcher, when he used a new technique to identify bits of human DNA that probably code for parts of genes – dubbed **expressed sequence tags** (ESTs). What's more, in 1991 NIH filed a patent application to cover more than 7000 of these ESTs, despite the fact that nothing was really known about them, apart from the fact that they were parts of genes switched on in the human brain. Watson dismissed Venter's EST approach as "cream-skimming", and called the patenting application "sheer lunacy", since "virtually any monkey" could do what Venter's group was doing. Under pressure from the scientific community, the US Patents Office eventually rejected the application, but the episode brought the whole issue of patenting life into sharp focus.

After the dust settled...

In the end, at the instigation of President **Bill Clinton** and British Prime Minister **Tony Blair** the race between Celera and the HGP was declared a tie, with Francis Collins and Craig Venter shaking hands and smiling for the world's press on June 26, 2000.

The public consortium then published its draft human genome sequence in the UK journal *Nature*, during the same week as Celera published its version in the US's *Science*, in February 2001. With hindsight, Celera's involvement probably speeded up the HGP, since the majority of it was finished two years ahead of schedule. The final version of the human genome sequence was eventually announced on April 14, 2003 by Francis Collins (Celera having dropped out after the draft stage). Since then, HGP scientists have been refining and annotating the sequence of each individual chromosome – the completion of the final and largest – chromosome 1 – was announced to an underwhelmed press in May 2006.

Perhaps inevitably, the hype that surrounded the announcement of the draft sequence in 2000 has led to some genome-fatigue, and several commentators have since questioned the benefits of such a large and expensive project. Others expressed disquiet over a perceived increase in genetic explanations for human nature. So the news that the human genome was thought to contain around 35,000 genes (more recently estimated at 25,000), rather than the 50,000–100,000 genes originally predicted, was gleefully reported by some as evidence that genes play a less significant role in both science and society than was previously claimed. But as researchers are finding out, it's not the number of genes that's important – it's how they are used, as well as when and where. After all, the round-worm *Caenorhabditis elegans* is about 1mm long, and made up of just 959 cells – yet its genome contains around 20,000 genes, nearly as many as humans. What accounts for the discrepancy? And, if only an estimated 1.5 percent of the human genome codes for genes, what is the remaining 98.5 percent doing?

It seems that the genes of more complex organisms, including humans, are adept at multitasking. Many can be "alternatively spliced" – that is, the sub-units that make up genes (called **exons**) can be stuck together in a multitude of different combinations (the genes of simpler organisms come as complete units, and so cannot be shuffled about in this way). Also, a particular gene may have several different roles, depending on which tissue it is switched on in, and at what point in development. There are a large number of control DNA sequences in the genome, associated with particular genes and responsible for fine-tuning their activity. These **regulatory sequences** probably make up around at least three percent of the genome.

It's estimated that about ten percent of human genes manufacture RNA molecules that are never used to make proteins. Many of these so-called **RNA genes** have roles in protein production – for example, the RNA that makes up the ribosomes and the adaptor tRNA molecules (see chapter one). There are also many pieces of human DNA code that make RNA molecules expected or shown to play a role in controlling gene activity – their numbers are difficult to pin down, since such sequences are hard to

Craig Venter (1946–)

The mainstay of the "private" effort to sequence the human genome (one of the sequences turned out to be his own) began as a scientist employed by the US National Institutes of Health. He quit the NIH in 1991 to start his own non-profit Institute for Genomic Research. But it was his move to found **Celera Genomics** in 1998 which brought real fame. A private corporation, Celera, and its flamboyant CEO, promised they would finish the human genome by 2001, four years before the (mainly US and British-funded) public genome project. And they were set to sell the results.

Venter has often ascribed his drive and ambition, which appeared after a youth dedicated largely to surfing, to his experiences as a medical orderly in Vietnam. He took up biochemistry on his return from the war, and worked in universities before joining NIH. He was an early user of the technique of making cDNAs from messenger RNAs, which he applied to the identification of genes which are active in the brain. Venter's attempt to overtake the official Human Genome Project's efforts was based on massive throughput "shotgun" sequencing, and he claimed to be meeting all his targets – though he did not publish the data, in order to protect its commercial value. But the controversy that followed made political waves that eventually led to the governments of the US and UK ruling that genome data should be made public. Celera's share price collapsed, and the triumphant joint announcement by the public and private efforts in June 2000 that the genome project was substantially complete effectively marked the end of the company's hopes of big returns from the data.

The competitive spur of Venter's project thus did more to speed up the overall effort than to privatize human genes. He was effectively fired by Celera in 2002, but after a short while he re-emerged in the public eye as chair of the J. Craig Venter Institute. His post-genome project interests include identifying the components of the "minimal genome" needed to support life, and – he has hinted – the synthesis of a living organism. He has also begun a survey of the genetic diversity of the living world (see chapter 5), and in 2005 he co-founded Synthetic Genomics, a biotech company hoping to produce biofuels from micro-organisms. And he is still trying to speed up human genome sequencing. In 2003 Venter promised $500,000 to the first people to sequence a human genome for $1000. Two years later he also joined the board of the X Prize Foundation, backed by the Canadian diamond company Archon, who in 2006 offered $10 million to the first private sector group to sequence 100 human genomes in ten days. Venter's contribution to the genome saga is well told in James Shreeve's 2004 book, *The Genome War: How Craig Venter Tried To Capture The Code Of Life And Save the World*. Shreeve had full access to his subject over a long period, and pretty much takes Venter's point of view over the disputes about DNA data. Venter has been working for some years on his own book about the implications of genetic information, drawing on his knowledge of his own genome, but this has yet to appear.

Francis Collins (1950–)

Francis Collins began his scientific career as a physical chemist but then qualified as a doctor, as a prelude to working in human genetics – first at Yale, then at the University of Michigan. He developed a reputation as a formidable "gene hunter", especially through his team's victory in a race between several labs trying to pin down the faulty gene responsible for cystic fibrosis. In 1989, Collins, Lap-Chee Tsui and John Riordan of Toronto's Hospital for Sick Children announced they had pinpointed the 250,000 base-pairs of the gene that is damaged in the majority of cystic fibrosis cases.

Collins was also closely involved with identification of several other disease-associated genes, including **Huntington's disease, neurofibromatosis** and one kind of **leukemia**. In 1992 the first head of the US genome project, James Watson, clashed with the powerful head of the National Institutes of Health, Bernadine Healey, who favoured patenting genes. Collins took over as director following Watson's resignation, and saw the project through to completion.

He remains director of the US National Human Genome Research Institute, developing strategies for comparative genomics for investigation of disease, and for unraveling the alterations in gene expression in cancers.

Collins has combined his new career as a big league science-bureaucrat with further work in the lab, extensive involvement in social issues around genetics, and work on stem cells. He has also been prominent in explaining how a serious scientist can be a devout Christian. As creationists – in their new guise as proponents of "intelligent design" – try to undermine the teaching of Darwinism in the US, he has taken pains to outline his view that evolution is the best account of the development of life on Earth. At the same time, he believes that the universe in which natural selection generates such wonders was created by God – a position which is at the core of his 2006 book *The Language Of God: A Scientist Presents Evidence For Belief*. The book devotes a good deal of space to the workings of the genome as an example of God's handiwork.

pick out. Recent years witnessed the discovery of tiny **microRNAs** which, along with their close cousins – small, interfering RNAs (**siRNAs**) – can curb or even halt the activity of larger mRNA molecules (the intermediaries between a gene and the protein it codes for). Their discovery suggests yet another important layer of genome control. The significance of this mechanism was recognized in 2006, when its US discoverers **Andrew Fire** and **Craig Mello** were awarded the Nobel prize for Medicine.

As well as protein-coding and RNA genes, the genome is littered with an estimated 20,000 pseudogenes – defunct genes which have, over time, acquired mutations and fallen into disuse. For example, around three hundred of the one thousand or so olfactory receptor genes that give mice and rats their sense of smell are now just pseudogenes in humans, reflecting our ever decreasing reliance on this ability throughout evolution. Intriguingly, it seems that some pseudogenes are still 'switched on' in certain tissues, suggesting they may not all be the relics everyone supposed them to be. Perhaps they too play a role in controlling the activity of other genes, as some preliminary experiments suggest.

What of the rest? At least forty percent of the human genome is made up of repeated DNA derived from mobile genetic elements called **transposons**, also called transposable elements (see p.99). Throughout evolution, these parasitic genetic elements have inserted themselves into host genomes where they seem to exist purely to get themselves copied. There are two sorts – **DNA transposons**, which move around by "cutting and pasting" themselves from one part of the genome to another, and **RNA transposons**, or **retrotransposons**, which act by "copying and pasting", resulting in the original copy plus a new one elsewhere. Retrotransposons are thought to have evolved from viruses that inserted themselves into the genomes of our ancient ancestors. These viruses eventually lost their ability to make new infectious virus particles, but some were still able to copy their genetic code. Transposons are thought to have played a key role in evolution, by pasting themselves in front of genes, and changing the way in which they are switched on.

So, in addition to straightforward genes, the human genome contains **regulatory sequences**, **RNA genes**, tiny **RNA control sequences**, **pseudogenes** and ancient **genetic parasites**. As with other complex genomes, the rest of it is made up of large stretches of DNA sequences that do not appear to have been conserved during evolution, suggesting they don't code for anything crucial. Some of this consists of blocks of repeated sections of DNA – genetic "stutters" (unlike the transposon-derived repeats, which are interspersed throughout the genome). This "tandemly repeated DNA" includes **microsatellites**, the variable regions used in DNA fingerprinting and gene hunting. But the remainder is effectively "dark matter" at the moment, to use a cosmic analogy, in that we don't really know what, if anything, its purpose might be.

Post-genome research

The end of the HGP marked the beginning of a deeper understanding of how genes affect the human body in health and illness, knowledge that will eventually improve health care. But having the complete human genome sequence is a bit like buying yourself a French dictionary – it's an invaluable and impressive tool, but it won't teach you how to speak the language, or much about France. The starting point for post-genome research is that we all have an estimated 25,000 genes, which somehow code for around 300,000 different proteins. These genes, their protein products and the regulatory RNA molecules encoded by the genome interact with each other (and a myriad of non-genetic factors) in ways that we don't yet understand. Identifying all these biological molecules and their connections, plus the variations involved in health and disease is a huge task – several orders of magnitude bigger that the Human Genome Project itself.

One fruitful area of research has been to sequence the genomes of other living creatures. Comparing human DNA with that of other species is a good way of picking out genes, since genes are made up of DNA sequences that often remain largely unchanged throughout evolution. Also, understanding the function of animal genes can throw light on their human counterparts. Non-human genomes sequenced so far include that of *Saccharomyces cerevisiae* (a type of yeast), followed by the fruit fly, puffer fish, roundworm, mouse, rat, chicken, dog and chimp genomes. (For more on how researchers are using **comparative genomics** to shed light on evolution, and to identify some of the genetic differences that make us human, see chapter five.)

Many scientists feel that the key to understanding human health lies in studying all the proteins found in our bodies, known collectively as the **proteome**. The science of proteomics involves working out the role that each of these proteins plays in the body – finding out which tissues a protein is used in, whereabouts in the cell it is located, which other proteins it interacts with and what its physical structure is. Another very active area of "omic" research is **metabolomics** – the study of all the chemicals, or **metabolites**, made or used by a particular type of cell. Some metabolites are not around for very long at all – they may exist for just a few seconds, which makes studying them a challenge. There are also **transcriptomics**, in which DNA "chips", or **microarrays**, are being used to look at patterns of gene activity in different healthy and diseased tissues by registering patterns of synthesis of the messenger RNA that is used for protein manufacture.

Other projects aim to identify variations in the genome, the most common of which are known as **SNPs** (single nucleotide polymorphisms – see p.97). Humans share 99.9 percent of their genetic information, which means that around 1 in every 1000 DNA base-pairs differs between any two people. Working out how these differences affect human health is another major challenge facing researchers: the **International HapMap**

Project is a large-scale collaborative effort attempting to catalogue all of the SNP variations in the human genome. There are other types of DNA variation too (see chapter four), and it remains to be seen how much each sort will affect a person's susceptibility to common illnesses. One way to try to find out how genetic differences affect our risk of developing conditions such as diabetes, heart disease, Alzheimer's disease and schizophrenia is to collect DNA samples and medical information from large numbers of people, and then do lots of statistical analyses. This is the aim of the UK's Biobank, and many other similar projects taking place around the world. Biobank (www.ukbiobank.ac.uk), launched in March 2006, eventually wants to recruit up to 500,000 volunteers aged between 40 and 69 years.

Another big project aiming to make sense of human genome data is the **International Human Epigenome Project** (IHEP), which is targeting the "control switches" that regulate genes. Specifically, it will look for positions in the DNA code where "methyl" groups are attached – chemical tags that act, via proteins, to shut genes down.

This research ought to shed light on how different body cells can produce different sets of proteins, despite having identical genomes. Understanding how genes are regulated could also provide some of the missing links between genetic and non-genetic factors (such as diet), and their roles in disease.

Biological databases

Biology, and molecular genetics in particular, is fast becoming all about information. It's estimated that the floods of data now being generated by the life sciences are doubling our knowledge in this area about every *six months*. Fortunately, because of the concurrent exponential rise of information technology in the past twenty years, there are ever more powerful computers available to handle all this information and specialists in **bioinfomatics** to work out how to do it. At the moment, there are over one thousand public and commercial biological databases in existence.

The first of these databases were set up in the 1970s, to store **protein sequences** (the order and category of a protein's constituent amino acid "building blocks"). Having information such as this stored centrally means that scientists can cross-reference their discoveries and find out, perhaps, whether someone else has already found a different member of the same protein family; or a protein that shares a common domain (subunits with similar shapes and similar functions) or the same protein but in a different species. Once DNA sequencing technology took off, **nucleic acid sequence** databases sprung up, allowing researchers to ask similar questions about genes, parts of genes and even just sections of DNA code. As of April 2006, over 130 billion base-pairs of sequence had been deposited in the GenBank (www.ncbi.nlm.nih.gov/Genbank) and RefSeq (www.ncbi.nlm.nih.gov/RefSeq) databases alone.

Now there are databases containing information on the human genome, various animal genomes, pathogen genomes, human genetic variation, genetic alterations associated with cancer, protein three-dimensional structure, genes linked to obesity, genetic variations linked to drug response and gene activity in different tissues – to name just a few. The **US National Office of Public Health Genomics** maintains a fairly comprehensive list, which can be found at www.cdc.gov/genomics/links/database.htm. There are also "meta-databases" that act as portals to many other biological databases, the best known of which is probably **Entrez**, hosted by the US National Center for Biotechnology Information (www.ncbi.nlm.nih.gov/Entrez). Here, you can search for a gene sequence, the protein it codes for, any diseases it may be involved in, and any publications it may have featured in.

There are also **genome browsers** (Ensembl, for example: www.ensembl.org/index.html) that allow scientists to visualize entire genomes, chromosome by chromosome, with all their associated biological data – a gene hunter's dream come true. With all this information at their disposal, many of the genes involved in single-gene disorders have now been found, and hunts for new ones can often be completed in a matter of months, rather than years. But the challenge now facing geneticists is to identify the role that genes play in **complex diseases**, which result from the interaction of many different genetic and non-genetic factors – a much harder nut to crack. This is part of the larger effort to work out how these genome databases relate to the others which will archive protein, metabolic, or RNA transcription data, from all the new "omic" studies. The dream is to integrate all these different kinds of data (and there is a vast and increasing amount of it) into a new "systems" biology. This would encode the rules that describe how all these different elements of living organisms interact. There are models for this kind of analysis in some simple organisms – at least in parts of their operation. But what a fully developed systems biology might look like, or how long it might take to figure out, are far from clear.

Genes ain't what they used to be

So far, we have been laying out some of the basic ideas about genes and DNA with a fairly simple take on the history included to knit the story together. Gregor Mendel proposed the laws of heredity, which were based on genes, that were then mapped, and found to be made of DNA. Eventually, scientists began to see more and more of the big picture, and tackled entire genomes – all the genes, and their DNA base sequences, which are packaged in the chromosomes of a particular organism.

However, ideas about what a gene is have become increasingly complicated lately. A gene was always a pretty elusive thing anyway, existing on a minute scale we cannot really visualize, in a world of cells, chromosomes and enzymes that are similarly mysterious. Lately even biologists have found it hard to agree on a clear definition of a gene. We know an amazing amount about them, but that doesn't always make it easier to say just what a gene is. DNA, it is understood, was always there waiting to be discovered. But genes are a bit more slippery. What they are is to a certain extent a function of what we know at any given time, and how we choose to define them.

The word is still useful, however. Science historian **Evelyn Fox Keller**, in her book *The Century Of The Gene* published in 2000, asked if the difficulties in saying just what a gene is might lead to the term being abandoned. She thought not, for three reasons. Talking about genes is still a convenient shorthand for scientists doing lots of different kinds of experiments. It points to particular things you can do to make things happen in science, industry, medicine or agriculture. We have learnt that genes are powerful, and that this power came from the things that genes were trying to explain, and what they must be able to do – somehow – for those explanations to work.

There are two key ideas here, which it is helpful to keep separate. One derives from thinking about heredity and resemblance. This was the problem that fascinated humans for thousands of years: how are some features preserved from one generation to the next? It is beautifully put in **Thomas Hardy**'s short poem *Heredity*.

I am the family face;
Flesh perishes, I live on,
Projecting trait and trace
Through time to times anon,
And leaping from place to place
Over oblivion.

Less poetically, the German biologist **August Weissman** asked in 1885, "How is it that … a single cell can reproduce the tout ensemble of the parent with all the faithfulness of a portrait?" The question, of course, applies not just to families but species – which can remain unchanged over hundreds of millions of years. Yet people looking at both species and families are equally fascinated by variation. We are like our parents, yet unlike them. Humans are all alike, but different. The answer to this apparent paradox of resemblance and difference turned out to lie in genes. The ground for the idea was laid by Mendel, with his hereditary factors, and by Weissman, who proposed that each organism harboured a "**germ plasm**" which was passed unchanged to the next generation. When Mendel's results were rediscovered in 1900, the germ plasm – now understood as a container of chromosomes – was soon redescribed as being the store of these factors. Factors which, in 1909, were named genes.

At first, the whole point of the word was that nobody had any idea what genes were. The term was invented by the Danish biologist **Wilhelm Johanssen**, who said that "the word 'gene' is completely free from any hypotheses". As he went on, less helpfully, "it expresses only the evident fact that, in any case, many characteristics of an organism are specified in the gametes by means of special conditions, foundations, and determiners which are present in unique, separate, and thereby independent ways – in short, precisely what we wish to call genes". The word caught on, and as we have said, it was found that genes could be mapped, plotted and mutated. Then the genetic substance was identified among the molecules of the cell. The DNA structure was solved. It fitted neatly into a story of resemblance and difference. But as time went on it became clear that DNA was far from stable – it had to be kept in good shape by a host of error correction and repair machines in the cell. The whole set of genes came to be seen, on the one hand, as an information store, an archive. On the other, it gradually emerged as part of a constantly managed and monitored system of the cell. By 1983, the veteran plant geneticist **Barbara McClintock** could write, in her Nobel prize-winning lecture, of the whole genome as "a highly sensitive organ of the cell, monitoring genomic activities and correcting common errors, sensing the unusual and unexpected events, and responding to them". Genes, it seems, are now simultaneously abstract ideas, physical entities, and in some ways also living things, as well as being strings of code! No wonder genes can be confusing to experts as well as the rest of us.

Spot the gene

There are nowadays a lot of complicated ways to "be a gene". For example, there is a stretch of DNA bases that contains the information for making two obscurely named substances found in human cells. One – known as INK4 – is a small protein involved in regulating the cell cycle and links with the origin of some cancers. The other is called ARF (which stands for ADP ribosylation factor) and it is part of another cellular switching system. It is involved with the manufacture of the small membrane-bound globs called **vesicles**.

Everything starts, as usual, with a DNA sequence that is transcribed into RNA. The RNA sequence contains four separated regions, or **exons** (see p.78) which will be used to code for lengths of amino-acid chain in a protein. But they are used to generate two different messenger RNAs for translation. The INK4 messenger is made by combining exon 1 with exons 3 and 4, while ARF derives from exon 2 with the same pair. But these are not just modules that are arranged in a different combination. The exon 3 and 4 unit is "read" starting from a different point depending on whether it is joined to exon 1 or exon 2. This creates a different reading frame for the three-bases-at-a-time amino acid code. So the two different proteins that can be made end up with very few amino acids in common.

None of this violates any of the rules of genes and coding as they were teased out in the 1960s. Nobody imagined then, however, that things would get quite so complex. Do we have one gene here, or two? Well, you decide. The existing conventions don't give any clear answer.

The true character of the gene?

The philosopher of biology, **Lenny Moss**, has a useful take on this story in his interesting but demanding book *What Genes Can't Do*. For him, the trouble with the gene concept began early in the DNA era. What he terms the "**classical molecular gene concept**" tied structure and function together. A gene, as we have described in earlier chapters, was a linear stretch of bases in DNA, which corresponded to a linear sequence of links in a protein chain. This is an immensely powerful idea, and the cornerstone of modern molecular genetics. Much of this book rests on it. But Moss argues that it also embodied a great confusion about what genes do. It is actually two ideas in one, and they should be kept separate.

The first, older notion could be called the *gene as character*. It has a whiff of preformationism, the old idea that the egg contains a tiny version of the adult organism, and marks an association with a **phenotype**, the actual set of characteristics the developed organism ends up with. This was where Mendel began, with his tall or short plants and differently coloured flowers. The other concept, call it *gene as information*, is the modern idea of a molecular sequence, as previously described.

The thing to understand is that this is a quite different kind of gene, or a quite different idea of what a gene is. It does not fix a phenotype. It is a kind of resource which a cell can use in different ways (or alternatively, not use at all). But some time around the invention of the **molecular gene** concept (that a gene was a linear stretch of bases in DNA that corresponded to a linear sequence of amino acids in a particular protein) the two ideas got welded together. And every news headline we now read concerning genes "for" all those complex traits ensures that the link stays strong.

In the DNA era this confusion just seemed to be the way things went in studies of the cell. The experimenters were busy figuring out how genetic information dictates the structure of a protein. And, in the simple viruses and slightly more complicated bacteria they were using, that was often what happened. But it is a short, though misguided, step from there to supposing that the gene dictates all manner of other things. And describing DNA as the master molecule, the genome as the Book of Man, or as the human blueprint, perpetuates the confusion. It ties up gene as *character* with gene as *information* in a peculiar blend of the old and new, a marriage of **preformationism** (that seventeenth-century theory that all the organs of an adult were prefigured in miniature within either the sperm or the ovum) and molecular biology.

In the 21st century, however, it is harder and harder to find genes that determine even a unique protein sequence. The classical molecular concept of "one gene, one enzyme", and **Francis Crick**'s central dogma that genes make RNA which make protein, began to unravel with the discovery of the enzyme **reverse transcriptase** in the 1970s. This is a DNA polymerase enzyme that transcribes single-stranded RNA into double-stranded DNA: it does the opposite of normal transcription, as it synthesizes DNA from RNA. The discovery allowed scientists to see DNA being made on an RNA template. Not long after came the revelation that genes in complex organisms are actually split into lots of bits and pieces, separated by stretches of DNA which are not part of the gene proper. And now we know that chunks of DNA can move around the genome, and that their RNA transcripts can be edited, spliced or even reordered before they get near producing a protein.

We still think genes are useful, however. There *are* pieces of information that are passed on from one generation to the next. And they tend to come together in units which work as one. Quite what those units are made up of can vary a lot, as we have seen, but they still need a name. The initial definitions may need qualification and expansion, but they still more or less work. One of the most recent, fattest, and glossiest textbooks on genomes (by Terry Brown, helpfully titled *Genomes*) still defines the word gene as "a DNA segment containing biological information and hence coding for an RNA and/or polypeptide molecule". Amid all the variation, and all the different ways the DNA "segment" can be built, those seem to be the essentials.

But these details are extraordinarily variable. A good way of summing up this character of the gene is to see it as a shape-shifter, as the anthropologist **Margaret Lock** has called it. It appears differently depending on the context, and depending on how you look at it. And that is true not just for different vantage points in biology. Out in the wider world genes "look different" depending on whether you a concerned mother or a research scientist, a patent agent or an actuary, a drug researcher or a psychologist, an environmentalist or a farmer, a philosopher or a policeman. We will be bringing in more of those points of view in the rest of the book. But in the end, they are all concerned with the meaning of the many messages written in DNA.

Genes evolving

The living world

Genes are often depicted as a static, unchanging store of information – like books, CDs or DVDs. These analogies are accurate in that they do explain how a cell uses genetic information. But more recent science has emphasized an increasingly dynamic view of the genome. As we have seen, genes are constantly switched on and off, tagged, labelled, and relabelled, as the organism grows and develops.

What is normally preserved intact, however, is the total package of genetic information that is passed on to the next generation. Genes are what we inherit. But even here, there is change. In fact, when we look at genetic information in the context of evolution, as we will do in this chapter, we need to think of genomes, too, as highly flexible, dynamic entities. Over time, genes can mutate, be reshuffled, switched on or off at different stages of development, or even be ditched altogether to meet the demands of an ever-changing environment. Without DNA's capacity for accumulating heritable changes, life on Earth would never have evolved into the dazzling array of plants, animals and microbes present today. Using modern DNA techniques, scientists are still filling in the details of Charles Darwin's famous theory, some 150 years after it was first proposed.

Natural selection in a nutshell

The fundamentals of evolution by natural selection can be stated very simply. This powerful idea does not depend on genes. It does need three elements: a population of things which reproduce; a way in which that reproduction can vary, so some of the new generation are different from the old; and a link between at least some of those variations and the chances of an individual in the population contributing to the next generation. In short: reproduction, variation and selection.

Nowadays that general idea can be applied to many things, from lines of computer code to, potentially, whole universes. But it is most developed in the modern view of the living world as a collection of creatures bearing heritable genes. Occasionally, random chang-

Charles Darwin, as photographed by Julia Margaret Cameron

es (**mutations**) occur in an organism's genetic material as it is copied from one generation to the next (see p.25). Most of these mutations will be anodyne, but some will affect the appearance, behaviour, or inner workings of the organism's offspring. Sometimes, an animal or plant with a particular mutation will survive better than its fellows of the same species living in the same place.

This process is the "**survival of the fittest**", to use the phrase coined by the Victorian writer **Herbert Spencer**. All it really refers to is how well an organism is adapted to its environment – how easily it can find and consume any available food, how well it can escape predators, and so on. And any advantage of this kind also increases the chances the organism will survive long enough to have offspring – which will all inherit the new mutation, and in turn pass it on to *their* descendants. In this way, mutations that have a beneficial effect, in this narrow sense, tend to become more common in a population, while harmful mutations will gradually become less common. Over many generations, natural selection results in the evolution of different species, each suited to its own particular environment.

This basically sums up the conclusions of **Charles Darwin**. His revolutionary ideas were based on the outward effects of mutations, of course, as his work was long before the discovery of DNA. But, for many of Darwin's contemporaries, he finally provided a vision of evolution backed by enough evidence to offer a convincing challenge to the biblical version

Charles Darwin and after

Charles Robert Darwin was born in 1809 in Shrewsbury, England, the grandson of the doctor, poet, and early enthusiast for evolution Erasmus Darwin. He began his academic career studying medicine at Edinburgh University, but when the tender-hearted young Charles realized that he couldn't really cope with the sight of surgery, then still performed without anaesthesia, he transferred to Cambridge to study theology. While there, he developed a fascination with natural history – the painstaking collection and identification of different species of plants and animals. His close friendship with a botany professor named **John Henslow** secured a place for Darwin on a planned two-year coastal survey of South America. The 26-year-old captain, Robert Fitzroy, required "a well-bred gentleman" to share his table, preferably a naturalist who could make the most of the opportunities afforded by the trip. So at the tender age of 22, Darwin set sail on the voyage that was to define his life and work, a journey that ended up lasting five years, and taking in Australia and the southern islands as well as South America.

When he returned, Darwin spent over twenty years analyzing his findings, mainly from his new family home at Downe in Kent, talking to animal breeders and reading many scientific articles in an attempt to make sense of "that mystery of mysteries", the origin of the species. In the early 1840s, a fellow scientist named **Alfred Russell Wallace** began corresponding with Darwin about evolution. But at that point Wallace had little evidence to back up his ideas, whereas Darwin had gathered a mountain of new data from other naturalists and from observations and experiments at Downe and in the surrounding countryside. He was working slowly on a book bringing it all together when, in June 1858, Wallace asked Darwin to review a paper he had written, entitled *On The*

of creation, in which God created all life on Earth in its present, "perfect" form. Sailing around the world for five years on HMS *Beagle*, Darwin was struck by nature's diversity, and by how many creatures seemed ideally suited to their habitats. For example, he brought back specimens of finches living on different islands in the Galapagos, and noticed later that they had different types of beak, depending on whether the food available was seeds, insects or both. Darwin, inspired by Malthus' famous essay on human population, saw life as a struggle for existence, arising from organisms' tendency to reproduce far faster than their environment could support. Individuals constantly compete for resources, to enable them to survive and reproduce. Darwin saw the many natural differences between organisms that make some better suited to their environment than others. He published his ideas in the book *On The Origin Of Species By Means Of Natural Selection* in 1859.

Tendency Of Varieties To Depart Indefinitely From The Original Type. This spurred Darwin into action, and Wallace's essay, along with excerpts from a paper written by Darwin some fourteen years earlier, were presented to the Linnean society in London less than a month later. In 1859, Darwin eventually published his famous book – the first run of 1200 copies sold out on the first day.

He went on developing his work until his death in 1882, notably in *The Descent Of Man* and *Selection In Relation To Sex* in 1872, and *The Expression Of The Emotions In Man And Animals*, as well as in books about orchids, barnacles and earthworms. (Although this Rough Guide is not a travel book, we urge anyone who can to visit Down House in Kent, now restored by English Heritage to look just as it did when Darwin researched and wrote there.)

By the end of the nineteenth century, most serious scientists accepted the fact of evolution. They were less sure about Darwin's mechanism of natural selection, but the advent of genetics eventually made it clear how it could work. The theoretical details were worked out in the 1930s by **Ronald Fisher**, **John Burdon Sanderson Haldane** and **Sewall Wright**, who united evolutionary theory with genetics in a framework known in the trade as the "modern synthesis". That theory stands today as biology's "theory of everything", as Dobzhansky claimed.

There are more biographies of Darwin than any scientist except Einstein. The best is probably Janet Browne's wonderfully readable two-volume treatment – Volume one: *Voyaging* follows the *Beagle* voyage and the conception of the theory. Volume two: *The Power Of Place* shows how Darwin wrote the *Origin Of Species*, and what came after.

Darwin also recognized that just as natural selection acts upon characteristics that affect survival, **sexual selection** can account for the evolution of features preferred by the opposite sex – some of which may seem, on the surface, to be irrelevant or even potentially disadvantageous. Traits such as the elaborate tail feathers of a male peacock, for example, seem to exist purely because female peacocks are impressed by them, and so are more likely to mate with males that possess them. So an inherited characteristic is more likely to get passed on to the next generation if it helps the organism survive to reproductive age, or alternatively, if it helps attract a mate.

Darwin's basic theory of evolution remains unchanged to this day, although details such as the relative importance of different forms of selection and evolutionary timescales have been hotly debated. Those Galapagos finches, for examples, are still evolving, and have been studied over recent decades by the husband-and-wife team of **Peter and Rosemary Grant**, as described beautifully in Jonathan Weiner's book *The Beak Of The Finch*. But there are countless other examples of evolution in action, too. Its scientific importance prompted Russian geneticist **Theodosius Dobzhansky** to write that "nothing in biology makes sense except in the light of evolution", the title of his 1973 anti-creationist essay. Along with Gregor Mendel's rules governing *how* inherited characteristics are passed on from one generation to the next (see chapter two), and work carried out in the first half of the twentieth century showing *what* is inherited (see chapter one), Darwin's theory still underpins all of modern genetics.

A very brief history of (human) evolution

The planet we inhabit is thought to be around 4.5 billion years old. The precursors of all life on Earth first appeared some four billion years ago, in the form of **self-replicating chemicals** (probably RNA – ribonucleic acid – a close chemical relative of DNA). The details are still hazy, but it is possible that it happened in the "warm little pond" of chemicals that Darwin conjectured. Over hundreds of millions of years, these chemicals gradually gave rise to simple microbes, then larger single-celled organisms with specialized internal compartments, followed by ever more complex multi-celled creatures. Some of these early life forms evolved into simple fish-like animals, which in turn – about 300 million years ago – gave rise to creatures that were able to live on land. The descendants of these early amphibians included dinosaurs, birds and all mammals, including the apes.

We know from the fossil record that all the species alive today represent just a fraction of those that have ever lived on Earth. The history of life on our planet is littered with extinctions. Some species made gradual exits while others were wiped out en masse – in the case of the dinosaurs, by sudden climate change, probably triggered by a huge asteroid collision. It was this chance happening, about 65 million years ago, that allowed the first rat-like mammals to evolve and diversify, with the first apes eventually appearing some 25 million years later. Our closest living animal relatives are chimpanzees – the last common ancestor shared by chimps and humans was swinging through the trees a mere 6 million years ago. Following the decoding of the entire chimp genome in 2005, it seems that chimps share 94 percent of their genetic material with humans. And the DNA that makes up the actual genes of the two species is almost 99 percent identical, making us as closely related as horses and zebras.

Our early ancestors, **Homo erectus**, first appeared in Africa 1–2 million years ago. They spread throughout the world and evolved into different species of ancient humans. Most scientists now think that our own particular species, **Homo sapiens**, descended from people who lived in Africa 150–200 thousand years ago. These first "modern" humans also spread around the world, eventually replacing all the other types of humans, including Neanderthals. So we are newcomers on the planet, relatively speaking. If the history of life on Earth was condensed into one year, so that each day represented about 11 million years, then the first backboned animals crawled out of the sea sometime during November, while our modern human ancestors appeared during the last half an hour of December 31.

On this scale, modern humans started farming about one minute before midnight – ten thousand years ago – leading to a sharp increase in their numbers. The idea that plants could be cultivated rather than gathered, while animals could be domesticated instead of hunted down, soon spread. Our human family is now growing ever more rapidly, following industrialization and medical developments such as antibiotics and vaccination. In 1800, about one billion people lived in the world. This number had doubled by 1930, tripled by 1960, and had officially reached six billion in October 1999. Because of this population explosion, everyone alive today shares recent common ancestors: the average marriage in Europe is between sixth cousins, who share a great, great, great, great, great grandparent.

Evidence of human evolution

Our bodies still bear witness to our ancestors' lifestyle – we have an appendix and wisdom teeth that we don't need, for example – but early humans required them to cope with the raw plants in their diet. This is also true of some human behaviour: an innate fear of snakes may be an inheritance from our early ancestors. Our history is also written in our genes. The types and numbers of variant genes, or **alleles** (see p.42), in a population can provide clues about the lives of our ancestors: their diet and the diseases they encountered. For example, in some parts of the world, many adults cannot drink large amounts of milk, because they are unable to digest lactose (the sugar found in milk). Drinking fresh milk causes them discomfort and diarrhoea, although they can eat dairy products treated to break down lactose – like yoghurt. This inability results from a shortage of the enzyme **lactase**, which breaks down milk sugar into simpler forms that can be absorbed into the bloodstream. It is normally present in babies until they are weaned, but production then shuts down. Elsewhere – in the majority of Northern Europe, for instance – people have a gene variant that allows them to digest lactose in adulthood, a characteristic that probably spread through some populations as their ancestors began farming cattle. Indeed, very recent findings suggest that several separate mutations in the same gene increased in frequency in different groups, as recently as 3000–7000 years ago.

Another example of recent human evolution is the high frequency of the **sickle cell** blood disorders and **cystic fibrosis** (CF), a disease that causes lung and digestion problems in some human populations. Both are serious genetic conditions that affect people who inherit two copies (one from each parent) of a disease allele. As the number of humans living together expanded, diseases like typhoid, cholera and malaria thrived, killing many of them. But people who inherit just *one* copy of a sickle cell allele are resistant to malaria, which is caused by a parasite that invades red blood cells. Similarly, unaffected carriers of one of the many CF alleles appear to be resistant to the effects of the microbes that cause cholera and typhoid. Since these particular genetic variations increased our ancestors' chances of surviving infectious diseases, they reached high frequencies in certain populations. Unfortunately, although the original threat has receded in many places, many of us still carry these protective alleles, increasing the chances that two carriers will meet and have a child affected by either CF or sickle cell anaemia.

Founder effects and inbreeding

Some alleles spread through a population because the people that have them are more likely to survive, and to pass those alleles on to future generations. Other alleles have no particular beneficial effect, but may become common in a population simply by chance – the phenomenon known as **genetic drift**. Some alleles, even harmful ones, can become widespread due to a **founder effect**, in which a community arises from a small number of individuals. The founder effect is thought to explain the high prevalence of particular genetic diseases in certain populations – such as the unusually high number of cases of a very rare condition called **porphyria variegata** in the Afrikaans-speaking people of South Africa. The features of this disorder vary widely, but can include acute attacks of stomach pain and extreme sensitivity to the sun. In 1972, a doctor named

Geoffrey Dean estimated that about 8000 South Africans affected by porphyria had inherited it from either Gerrit Jansz, a Dutch settler in the Cape, or his wife, Ariaantje Jacobs. Dean also estimated that 1 million of the 3 million whites living in South Africa in 1972 were descendants of forty original settlers and their wives, making the high incidence of porphyria a likely textbook example of the founder effect.

Other examples exist in most communities that have remained relatively small, and isolated either geographically or culturally, like the **Amish population** of North America. Even apparently unrelated members of such groups will share several harmful alleles, some of which will only cause disease in people who inherit a double dose (one recessive allele from each parent). This is true of close relatives in any community, isolated or not, with the result that incest is a cultural taboo in most societies – the more closely related two people are, the more likely they are to carry copies of the same disease alleles.

Marriages between cousins are allowed in most countries, and in fact the risk of cousins having a child with a genetic problem is not that much higher than two apparently unrelated individuals, because we are all descended from so few people, relatively speaking. Unrelated parents have around a two percent risk of having a child with a genetic problem, while in first cousins this risk rises to five percent. So on average, 95 percent of the children of first cousins will be healthy, although the risk of a recessive disorder increases if there is a family tradition of such marriages. In addition, certain disorders are more common in some populations – for example, cystic fibrosis in people of white European origin (for reasons explained above) and thalassaemia in people of Asian and Mediterranean descent.

The spice of life

The raw material of evolution is **genetic variation** – without it, the process of natural selection would have nothing to act upon. There are lots of different types of DNA variation, or mutations, which can affect an organism in a variety of ways. The most common type involves just a single base-pair (or letter) of DNA. These are properly called single nucleotide polymorphisms, or **SNPs** – pronounced "snips" – which may have either a neutral, harmful or beneficial effect. If a SNP occurs in a gene, then there will be two alternate versions of that gene, or alleles, in the population. Any particular gene may come in several different versions, which may be inherited in any combination.

An example of the effects of different alleles is provided by the ABO blood grouping system, first described by Austrian doctor **Karl Landsteiner** in 1900, but only fully explained at the DNA level in the early 1990s. The ABO system is based on two different antigens (cellular "tags"), A and B, which are found on the surface of red blood cells. You may have one, both, or neither of these antigens. Group AB people, the least common in the UK, have both. Group O, the most common, have neither. Knowing your blood type is of crucial medical importance, since if you were to receive blood that contained an unfamiliar antigen, antibodies in your own blood would clump the donated cells together and you could die. Which blood group you belong to depends on which versions of the gene responsible for making these antigens you have – which in turn is determined by a handful of different SNPs. If you inherit an A allele from each of your parents, or one A and one O, then you will be blood group A. Similarly, people with two Bs, or a B and an O, will be blood group B. Someone who gets an A allele from one parent and a B from the other will be group AB, while two O alleles make you group O. You can see why blood groups are a useful way to test paternity.

It's not just blood groups that make us different. After the decoding of the entire human genome in 2003, it was estimated that everyone differs by around one SNP in every thousand base-pairs of DNA code – in other words, they said, humans are 99.9 percent identical (or 0.1 percent different) at the genetic level. Some of these SNPs could affect our appearance, the way we react to certain medicines, or our predisposition to conditions like asthma or diabetes. Others may have no effect whatsoever, especially those that occur in the vast tracts of DNA that don't contain any genes.

In addition to SNPs, there are many variable bits of DNA known collectively as **INDELs**, which are locations where a few base-pairs of genetic code have been deleted or inserted. Again, whether they have an outward effect or not depends on whether they occur in a gene or its surrounding control regions, or in DNA that is not part of genes. In recent years yet another type of common genetic variation has come to light, involving much bigger chunks of DNA (between 1000 and 3,000,000 base-pairs, in fact) – dubbed "copy number variants", or CNVs. These variable sections of genetic code, some of which contain many genes, may be duplicated, missing or even flipped around – but what effect these differences have on a person's health or appearance is anyone's guess at the moment. One thing's for sure, people aren't as genetically similar as previously thought, since while the majority of genes are identical between individuals, the number of *copies* of a particular gene may vary widely. Recent studies

Barbara McClintock (1902–1992)

Born in Connecticut and raised in New York, Barbara McClintock studied genetics at Cornell University in the 1920s, and developed a penetrating approach to plant reproduction. Her work was a world away from the headline-grabbing molecular biology of the later twentieth century, but she anticipated some of its most startling findings about how genes are organized.

She was an unusually skilled experimenter, and made many discoveries about the chromosomes and genes of maize, which became the mainstay of plant genetics. Her eye for the particular characteristics of an individual plant – she followed in particular the patterns of colour in corn kernels – and her ability to visualize how its chromosomes behaved, was captured in the title of science historian Evelyn Keller's biography, *A Feeling For The Organism*.

In the 1940s and 1950s, McClintock's work on the cytogenetics (following events inside the cell) of maize led her to the radical idea that some genes can move around on their chromosome, or even between chromosomes. But her work attracted little attention. And although she eventually got a secure position at the Cold Spring Harbor Laboratory in New York State, she published little and became a somewhat marginal figure, although she went on doing brilliant investigations of her chosen plants.

The importance of her transposable elements, as they were called, finally became apparent to everyone else after the discovery of the same thing in bacteria, where "jumping genes" were revealed in the 1970s. She received many awards, and finally collected her Nobel prize in 1983, for work mostly done nearly forty years earlier.

Her earlier findings undoubtedly failed to make the impression we now know they deserve. There were several reasons. There was prejudice against women in science, although she was the third woman to be elected to the US National Academy of Science, in 1944, and some of her professional disputes were with women. The fact that she did traditional genetics on plants, and never worked on molecular biology made her work seem impenetrable to many younger scientists in the 1950s and 1960s. Indeed, James Watson, who became director of Cold Spring Harbor, regarded her with something like contempt. However, her own somewhat unforgiving temperament probably had something to do with it, too. As one of her subsequent biographers put it, "at least some of her professional difficulties arose not from the fact that she was a woman, but from the fact that she was Barbara McClintock".

suggest that while such variation is normal, many common diseases may be triggered by the loss or gain of copies of certain key genes.

Evolutionary short cuts

There is only so much that geneticists can infer about evolution and natural selection by looking at populations and differences between individuals. To get to the real nitty-gritty of how evolutionary change comes about requires a close look at DNA, something that has only been possible during the last twenty years. In particular, the possibility of comparing whole **genomes** (the entire genetic code) of different organisms, only effected in the last few years, has led to some intriguing insights into how different species might have evolved.

One potential problem with explaining the process of evolution is the length of time it would take if it involved changing just one or two DNA base-pairs at a time. This works fine for bacteria that can reproduce roughly every twenty minutes, but is not such a great model for creatures that take several months to produce offspring. Furthermore, a single mutation in a gene is just as likely to cause harm as it is to have a beneficial effect. So if humans had evolved in this way, it ought to have taken a lot longer for us to get to where we are today. This stumbling block has been partly explained by the discovery of genetic "**copying and pasting**" – scientists estimate that an average gene in a complex life form is duplicated around once every 100 million years. Duplicating a gene allows one of the copies to acquire new mutations, which may or may not prove useful, while the other remains intact and able to fulfil its original purpose. Entire **gene families** have been identified, in which each member is thought to have arisen via the duplication and subsequent tweaking of one ancestral gene. Sometimes one of the gene copies is abandoned altogether – permanently switched off. The human genome is littered with the rusting hulks of inactive genes.

Evidence for more evolutionary short cuts has been unearthed from the study of individual genes, which shows that bits of DNA code found *within* a gene can also be duplicated several times. This is possible because, unlike bacteria, the genes of more complicated organisms are broken up into short sections called **exons** (see p.78). This peculiar feature is thought to have played a key role in evolution, as it stops organisms having to continually start from scratch when new proteins are needed. Duplicating exons within a gene can expand a useful part of the protein it codes for, or, as with whole gene copies, it can provide new raw material that can acquire mutations without affecting the existing protein too much.

Having genes broken up into functional modules also allows for shuffling and swapping around between different genes – like a giant Lego set, bits of one gene may be transported into another to make entirely new combinations. This process is believed to be triggered by a kind of genetic parasite called a **transposon** – a piece of DNA that is capable of "jumping" from one location to another within the genetic material of a cell. There are several different types of transposon, multiple copies of which can be found nestling in the genetic material of most living things. The first transposons were discovered in corn by **Barbara McClintock** in 1948, but few noticed her inexplicable finding in the pre-double helix era. She lived to see jumping genes identified in many other organisms, and was eventually awarded a Nobel prize in 1983. It seems that transposons have played a vital role in human gene evolution, illustrated by the recent finding that at least forty percent of the human genome appears to be made up of repetitive, transposon-derived DNA.

Finally, there is another way in which organisms can exploit existing genetic information in order to maximize its coding potential: by controlling gene activity. Some genes make proteins that control many other genes, so a mutation that ramps up or dampens down the activity of one of these control genes could have a widespread knock-on effect. Or a gene might be switched on at a particular stage in development, or in a particular body tissue, wherein it previously played no role.

Studying the genes of different organisms will eventually shed more light on which of the above mechanisms has been involved in the evolution of the many features and chemical processes responsible for everything alive today. This type of data is also being used in other, related fields of research, which we shall cover in the rest of this chapter.

Studying biodiversity

The theory of evolution helps explain how the staggering profusion of different species that the world is home to came about. As with many other areas of biology, study of DNA is revolutionizing the approach to their study. But how many species are really out there in the world, and which are related to which? The latest answers show that there are dimensions to life's variety – nowadays termed **biodiversity** – which were hardly suspected before. They are important both for their general interest and for the continued survival of the human race. It is astonishing how rich the biosphere is. But we are also in a race to comprehend the full extent and importance of biodiversity even as our own less commendable habits are depleting it.

You can recognize species by the old-fashioned method of just looking carefully, and noting the features of the creatures you find. Hunter-gatherer peoples are generally reckoned to be able to recognize 10,000 or so different animal and plant species in the rainforest. Evolutionary psychologists have pointed out that the survival advantages of being able to classify and name all those kinds may be one reason why our brains became so good at registering small differences in collections of similar things.

The Swede **Carl Linnaeus** (1707–1778) almost matched the hunter-gatherers by recognizing 9000 species. And he founded the modern system of classification, or taxonomy, for organizing and naming them all. Since then, the numbers have kept on rising, but the principles have stayed the same – give or take a microscope and a few measuring instruments. We know most of the birds and mammals – though occasional unknown ones still turn up. But there are a *lot* of other species out there. Shake a tropical tree – or better still fumigate its canopy – and the chances are that insects will fall beneath which have never been seen before. We have now seen at least one specimen of perhaps 1.8 million contemporary species. It is hard to be exact because there is not yet a central database, and some may have been "discovered" twice. It is even harder to be exact about how many different species there could be in total. Respectable estimates vary from 3 million to 30 million, with some arguing there could be as many as 100 million. If there are a million kinds of insect, and 400,000 different beetles, who knows how many microbes there might be?

DNA comes into this field in several different ways. They all depend on the specificity of genetic information, and the clever machinery that has been developed for fishing out sequences of genetic code and comparing them with what is already in the molecular data banks. If the sequence is already there, then that can give an accurate fix on which species the sample comes from. If it is not, chances are we have here a new species, or part of one.

Of course, relating a real world throbbing luxuriantly with life in virtually every corner of it to the orderly arrays of molecular data is more complicated than that. For one thing, the orderly arrays may not be there yet. Setting up such a database is the goal of the **Consortium for the Barcode of Life**, a collection of organizations who hope to compile DNA sequence information from each known species. The idea is to choose a short DNA sequence from a standard part of the genome which all organisms (or at least a large group of them) share. So you do not need the entire genome, just the sequence of one sub-unit of a key protein. The sequence chosen must be one where small variations within species are much less significant than differences between species. The candidate sequence for higher animals like mammals and reptiles, for example, is part of an essential protein in the **mitochondria**, just 648 base-pairs long.

There is a long way to go: getting the project funded properly, agreeing global standards for the procedures to follow, and of course the mammoth task of compiling all the data. The specimens can come from existing stores in museums, zoos, seed banks, culture collections and the like. Assuming the taxonomists, who love a good argument, agree as to what species the specimen belongs, the DNA barcode should soon be readable for a few dollars. Pilot projects in birds and fish are well under way, and plants and plankton will follow. And the major existing genome databases – Genbank and its siblings in Europe and Japan – already have established barcode sections.

The DNA-based system of species taxonomy should have plenty of advantages, its advocates claim. An unknown specimen can be identified from odd bits and pieces, and this will help to prevent illegal trafficking in protected species. This is a huge problem, for unscrupulous traders try to evade controls by denying their merchandise comes from particular species. The **US National Oceanic and Atmospheric Administration**, for example, recently prosecuted the dealer of a ton of dried shark fins, traded for soup, after DNA identification confirmed that the fins came from a protected shark species. The DNA method also works at any stage of the life cycle – a creature's form may change but the DNA stays the same. And it may become more accurate, faster and easier. The ultimate goal is an electronic hand-held field guide, a kind of "Hitchhiker's Guide to the Biosphere" which will read the DNA from a specimen and tell you where it comes from. That would be a great way to allow more people to get involved in biodiversity studies, but it is still some way off. For more information see: http://barcoding.si.edu/

Another approach to using the new DNA technologies to take samples of life forms does not depend on specific specimens. It need not even involve specific organisms. A number of projects focus on biodiversity by looking at the *range* of genes – on the grounds that evolution ultimately works one gene at a time. If a gene is found that is unknown or, more interestingly, does not seem to be even related to any of the genes already in the data banks, then perhaps it comes from an unknown form of life. So there are projects that work by taking an inventory of all of the DNA in a sample of, say, sea water or soil. Then the results are compared with DNA sequences from organisms whose genomes have already been studied.

One pioneer in this field was **Craig Venter**, who, as founder of Celera, headed the private sector's contribution to the Human Genome Project (see p.73). His ousting from the company not long after the project was completed did not dampen his appetite for sequencing DNA. He soon announced plans to catalogue all the genes on the planet. The key would be **"shotgun" sequencing**, which finds the order of the bases in small pieces of DNA derived from chopping up genes, and then tries to solve the resulting puzzle to recreate functional stretches of code. But Venter applies shotgun methods not to fragments from one organism, but to samples of whole ecosystems. This approach involves dealing with lots of genomes at once. Hence its name – **metagenomics** – is indicative of its going one better than standard genomics.

Venter initially combined business with pleasure: he sampled marine microbes whilst sailing his massive yacht *Sorcerer II* round the world. Never one to undervalue his own efforts, he compared his trip to Darwin's voyage on the *Beagle*, suggesting that he would inventory "what Darwin missed". It turned out that Darwin had missed quite a lot. Venter's first report in 2004 looked at the **Sargasso Sea**, regarded as an uninteresting environment, poor in nutrients. But when he fished for DNA he found 1800 new species and over 1.2 million new genes, as many as the total already known. These are all genes identified by computer analysis, so working out what they might all do is quite a different problem. But Venter remains committed to building a much more detailed picture of "who lives here". You can find out more about his vision at www.sorcerer2expedition.org

Venter has a knack for grabbing the headlines, but others are also working along the same lines. Notable is the **International Census of Marine Microbes** (ICoMM), a project of the Census of Marine Life, a ten-year project that began in 2000 and relies on 1700 researchers in over seventy countries. In July 2006, an ICoMM-affiliated group based at the US Woods Hole Marine Biology lab reported DNA analysis showing there were 20,000 kinds of bacteria in a litre of sea water, the vast majority of them never identified before. They were expecting to find just a few thousand.

Most of the new species were present in very small numbers, and ICoMM suggested that these and other samples – from locations in the Atlantic and Pacific oceans – demonstrate that there is an unmapped "rare biosphere". As the oceans are so vast, this data still represents a mere drop in the ocean (as it were) of microbial life. So there is much more species sampling to do in order to discover if other sites yield yet more rare varieties, or whether there are locations where some of them are more common. Current thinking is that they are an important reservoir of genetic diversity, but the role they play in ocean ecosystems remains a matter of speculation.

While there are untold numbers of microbes in the seas, they are not the richest environments on earth. A sterner test of the power of metagenomics, and of methods for making sense of the data gathered, will be **soil samples**. Craig Venter is starting work here, too, but microbiologists caution that the number of genes under our feet could be overwhelming. Processing his Sargasso Sea samples involved shotgun sequencing fragments adding up to a billion base-pairs. Covering the genomes in just one gram of soil probably involves dealing with 250 billion base-pairs.

Craig Venter

Don't expect a catalogue of new soil microbes any time soon, then. However, even microbes are not the bottom of the unplumbed depths of biodiversity. Another startling recent development involved the study of viruses. Their status as "life" may be debatable, because they need other organisms to reproduce. But these scraps of DNA, or sometimes RNA, can certainly contribute to genetic diversity. And it now looks as if viruses – specifically the kind known as **phages**, which prey on bacteria – harbour an immense and hitherto unsuspected reservoir of genes. According to a report in the science journal *Nature* in June 2006, "there may be more undiscovered genes in the viral world than in all other forms of life combined". Genetically, as the authors put it, we are froth on the surface of a viral ocean.

These unknown viruses, it seems, are found everywhere you look – in sea water, mud, and excrement. There are probably a thousand different viruses in your gut, and a million in a kilo of mud from the sea bed. And they are busy recombining DNA with each other, and with host cells, creating DNA sequences never seen before, including new genes. Most of this random shuffling of DNA creates rubbish. But with perhaps a trillion trillion new virus particles being created on the planet every second, useful new sequences are bound to appear.

The moonbank?

Preserving gene samples on Earth is all very well, but suppose our home planet suffered a catastrophe so complete that nothing was recoverable? The US-based **Alliance to Rescue Civilization** wants to cover that possibility by persuading backers to build a moonbase that will guard the essence of life and culture. As well as a really big (computerized) library, their project calls for a **DNA bank** with samples of all life on Earth. The facility would be run by people who, presumably, would also have to find time to repopulate the devastated Earth with humans.

The group's main inspiration, New York University biochemist **Robert Shapiro**, sees the moonbank as a sensible insurance policy, and a reason to

boost the space programme. NASA has declined to get involved so the ARC is a private outfit – so far restricted to mere advocacy and developing plans for such a base. But perhaps the multiple perils of the 21st century, from global warming to pandemics, and nuclear or bio-warfare, will strengthen their case – that, at any rate, is what Shapiro's colleague, science writer William Burrows, argues in his 2006 book *The Survival Imperative: Using Space To Protect Earth*. One question remains, however – is the moon far enough away? Perhaps they should be trying for a Marsbank?

Robert Shapiro

DNA and conservation

It is one of the environmental paradoxes of the early 21st century that new depths of biodiversity are finally coming into view, just as we are growing painfully aware that biodiversity of the old-fashioned kind – plants, insects, reptiles, birds, mammals – is being trashed. We may not know exactly how many species there are. But we do know that plenty are disappearing fast.

Trying to conserve what is left demands a mixture of political, economic and environmental action. But better knowledge of genetics can at least help plan the details of strategies to keep a particular population going. Failing that, conserving the genes themselves is better than nothing – though it is no substitute, of course, for having the living creatures out there that they belong to.

The importance of DNA to conservation can be seen from just a few examples. Suppose you want to help conserve grizzly bears in Montana. The population is poorly studied, and sampling grizzlies is a daunting prospect. But just as a few samples taken from the scene of a crime can reveal who was there, collecting bears' hair allows us to access their DNA. Analyzing the samples collected along survey routes and from "systematically positioned hair snag stations" will show how many bears there are, and how far they range.

Going into the DNA can also show how much genetic diversity there is *within* a population at risk. If the remaining population of a bird or a butterfly, say, is genetically uniform, there is a lower chance some individuals will survive a new environmental stress, such as a drought or a heatwave. And when intensive efforts focus on just a few creatures, like the **Californian condor**, individual DNA analysis can even be used to prevent inbreeding. In the condor's case, this has turned out to be necessary because some birds carry a genetic disease, which can be prevented if the right pairs are bred in captivity for later release.

This kind of work blends into the DNA bank option for gene conservation because, if things do not turn out well, the preserved samples may be the main component of the remaining gene pool. DNA collected from **black rhinos** in Zimbabwe a decade ago, for example, was originally intended to help study genetic variation in the population. But continuing onslaught from poachers means the DNA is all that is left of most of the animals.

But there is an argument for gene banking on a much larger scale. Some DNA banks already exist, usually based on existing collections from the pre-genomic era. Kew Gardens in the UK, for example, currently holds 22,000 samples of **plant DNA** from known specimens, which are available to researchers worldwide for a contribution to the shipping cost. New samples can also be extracted on request (Kew has a lot more than 22,000 plant species). The DNA storage goes along with existing banks of frozen seeds, embryos and pollen. Kew is also involved in the Millennium Seed Bank, which aims to preserve 24,000 plant species from all round the world. In the US, the genetic ark for plants is at the Missouri Botanical Garden, and there are smaller efforts being undertaken in Australia, South Africa, and Japan. In June 2006, Scandinavian heads of state met on the island of Svalbard in the Norwegian Arctic to launch the building of a "doomsday vault" for crop seeds, designed to survive future war or other disasters at lower latitudes.

In the animal realm, the "Frozen Zoo" in San Diego, which began a genetic bank of tissue samples in 1975, has material from over 7600 rare and endangered species, mainly from mammals. They include eggs and sperm, embryos and body cells, but all contain the vital DNA. Some of the later additions came from the last known survivors of a species now extinct, such as the **po'ouli bird**, which died in a Hawaii conservation centre in 2004. A selection of tissue samples was sent to San Diego, and the technicians there managed to establish a tissue culture from just one of them. (See cres.sandiego.org/projects/gr_frozen_zoo.html)

These are important projects in their own right, but they are also demonstrations of future possibilities. Over time, and with work on global coordination and universal standards, these banks will become part of a larger network, and will gradually merge with all the other DNA depositories and sequence records. We may not be able to preserve all the species we would like to, but at least their genes will still exist.

Mapping evolution

As well as giving us new insights into the variety of life that evolution has produced, new DNA technologies can provide fresh ways of studying evolution in action. Again, the most impressive thing tends to be the unprecedented detail of the observations, which is leading to startling insights into evolutionary mechanisms down among the molecules.

The most basic use of DNA sequence information is sorting out all the branches and twigs on the evolutionary tree. Comparing sequences from organisms which are alive today can show how close or how distant they are in evolutionary terms.

So one kind of DNA map is simply a map of relationships, of the kind which evolutionists have been drawing for many years. Such maps can be based on molecular data, either from analyzing proteins which two species may have in common, or directly from DNA sequences. Every living thing is, all the evidence suggests, descended from a single common ancestor. Any two kinds of creature living today will have a most recent common ancestor, their last close relative before the species diverged into the forms we see today. If the rate at which DNA changes is roughly constant, then comparing sequences from the two kinds is an indicator of how long ago their predecessors became separate species. That constant rate is a big assumption, but seems to fit other evidence for regions of genomes which can mutate away quietly free from selection pressures. This happens in some duplicated genes that have fallen out of use, in regions of "junk" DNA, or even in coding sequences that harbour silent mutations (that is, base changes which do not alter the amino acid in a protein).

Put this kind of information together with regular **fossil analysis** and close observation of other features of creatures alive today, and the tree of life can be filled out in enormous detail. The crucial point is that the DNA of the present-day creatures preserves information that has been around a very long time. (Most of it, anyway.) Life carries its own archive around with it. So we can use the present to reconstruct the past. We will say no more here about the details of the big picture of evolution which emerges – you can read an impressively complete account in **Richard Dawkins'** *The Ancestor's Tale*, as near to a systematic zoology as a popular science book ever gets.

Instead, let us look at new results about a species of particular interest, *Homo sapiens*. Early work tracking genetic variation before the advent of DNA settled some vexed questions. The first survey of human genetic variation was as long ago as 1919, and used the newly discovered **blood groups**. In fact, blood groups were a staple of **human population genetics** up to the 1950s, when they were supplemented by data on variations in sequences of a protein that every human has in vast amounts – **haemoglobin**. More sophisticated genetic analysis pioneered by **Richard Lewontin** at Harvard in the 1970s marked the final demise of the old belief that *Homo sapiens* originated several times in different regions, and that this accounted for the origin of different "races". The notion of race, throughout its dubious history, has emphasized features you can see – but these turn out to be truly superficial differences between humans. Lewontin found that roughly 85 percent of the genetic differences between humans are *within* populations. Another seven percent of the variation is bound up with differences between populations of the same "race" – between Southern and Northern Europe, for example. That leaves just eight percent of the total variation distinguishing the racial groups. The conclusion is that race is a culturally powerful notion, but biologically it's insignificant.

Now that has become an established truth, more recent work has concentrated upon the history of the one human race. The mapping, this time, is not of evolutionary relationships. It tracks migrations over much shorter timescales. Using genes to reconstruct history at this much finer resolution depends on two special elements of the human genome. As we explained in the last chapter, almost all our genes exist in two copies (see p.43), one on each paired chromosome, and these are shuffled more or less randomly when the reproductive cells copy their DNA. This muddles up the inheritance of genetic markers over time, and any pattern of past relationships gets blurred.

There are two exceptions to this. One is the separate gene complement of the **mitochondria**, the energy factories of the cell. These tiny bodies in the cells of higher organisms are descended from bacteria, and have kept their own DNA. When sperm meets egg, the father's mitochondria are lost – so everyone's mitochondrial genes, by default, come from a long line of female ancestors.

Similarly, most of the genes on the Y-chromosome – which belongs only to males – exist in just the one copy, and are inherited without the business of crossing over that involves all the other chromosomes. So we have two sets of DNA sequences with markers that can be used to track the past relationships between present-day populations: one for females, one for males.

The mitochondria have far fewer markers – Y is a small chromosome, but still has 50 million base-pairs, compared to the 16,000 of the mito-chondrial DNA – but mutate faster and have more copies of their DNA. This made them easier targets for the first analyses. The mitochondrial archive indicated clearly that all of *Homo sapiens* had a common ancestor in Africa as recently as 150,000 years ago: the individual dubbed "**mito-chondrial Eve**" by the press. The name is misleading, as she was neither the first human, nor the only one alive at the time. It just happens that the mitochondrial lineages founded by her sisters have all died out.

The Y-chromosome, with its comparative richness of markers, allows more refined analysis. The patterns of variation in the DNA that men carry in their Y-chromosomes today also indicate an African origin for *Homo sapiens*, though the oldest common ancestor this time is a mere 60,000 years ago. He will, of course, have had ancestors who were mito-chondrial Eve's contemporaries, but the variation in Y peters out before we get to them.

The Y-chromosome analysis has been developed in great detail, and has been used to reconstruct the movements of populations since the first modern humans began to trek out of Africa on their way to achieve our present global reach. This African exodus has now been mapped using genetic data, and cross-checked with experts in archeology and linguis-tics. The upshot is that we have a pretty good idea when humans first appeared in the various parts of the globe, and the paths our ancestors took to reach their ultimate destinations. The story is fascinating, and has been told in a number of excellent recent books. Best of the bunch is Spencer Wells' *The Journey Of Man: A Genetic Odyssey*. You can also follow his more recent analyses at the very fancy website at www3.national-geographic.com/genographic/journey.html, which also invites you to submit your own DNA sample to extend the database he uses, in return for a "map" of your own ancestors.

But there are more than two kinds of evolutionary map. The **International Haplotype Project** (HapMap) is going beyond the Human Genome Project by comparing genetic sequences from people all round the world to see what the differences are, and how they are distributed. The map begins with differences in individual base-pairs – the SNPs we mentioned earlier (see p.97). SNPs that are inherited together are then recorded as **haplotypes**, which are catalogued using SNPs that identify them as unique. The SNPs do not tell us anything about genes directly, but the idea is that spotting haplotypes that seem to more commonly accompany various diseases will offer new clues to where genes which influence the disease might be found.

Although the motive was medical, the first HapMap – finally published in 2005 – has other uses. One of the first was to spot which human genes are still evolving. People often ask if human evolution has stopped, or at least slowed down, because we are so adept at shaping environments to suit ourselves. There are general arguments that the answer is no, mainly to do with the continuous selection from the effects of disease, for example. The news from analyzing the HapMap is that quite a few genes have altered recently for other reasons, too.

Recent here means in evolutionary terms, of course – not last week. The evidence is from gene variants that have begun to spread through particular populations, but not become part of everyone's genetic make-up – like the lactose-digesting enzyme variants we discussed earlier. A team from the University of Chicago used a new way of processing the HapMap SNPs, one that shows if there are differences in the amount of overall genetic diversity between a group of individuals who have a new gene variant and those who do not. If the first group is less diverse overall, this is taken as a sign of recent selection. The team found evidence of changes in the last 10,000 years or so in genes instrumental in taste and smell, digestion, skin colour, bone structure and brain function. Like the HapMap published in 2005, this was just a first go at highlighting parts of the genome which are changing, so there is bound to be more to come along these lines.

Moving further back down the road that led to *Homo sapiens*, a different set of recent results shed light on another question thrown up by our first acquaintance with genomes that has puzzled many. How come, if our genes are so similar to the primates we already sense are our close evolutionary relatives, that we can do some things that are so different? Surely, the argument goes, the four percent (or so) sequence differences in the DNA of humans and chimpanzees cannot be nearly enough to explain why humans keep chimps in zoos, and write books about them, and not the other way round?

One solution to the problem, discussed for decades, could be that changes which set humans apart from chimps in the 6 million years since their latest common ancestor was alive were mainly in **gene regulation**, not in genes themselves. The detailed analysis of genomes and their expression that is now becoming routine finally seems to confirm this suspicion.

Yoav Gilad of the University of Chicago and his colleagues at Yale and in Australia used the new technology of **microarrays** (see p.54) to compare patterns of gene expression in humans, chimps, orang-utans and rhesus monkeys. The microarrays were specially prepared for each species, and each detected many messenger RNAs which matched particular DNA sequences. The results show how the relative abundances of mRNA, and thus the activity of particular genes, differ between the four kinds of primate.

They found that most genes are active at pretty much the same level. But, crucially, they also showed that a group of genes which were significantly more active in humans than in their primate cousins was coding not for any common or garden enzymes or structural proteins, but for **transcription factors**. In other words, they were genes which influence whether other genes are switched on or off. The implication is that crucial evolutionary changes behind human development involved *control systems* for already existing genes. As we keep discovering more about these – they involve a lot more than transcription factors (see p.30) – there is doubtless more news like this to come.

So the new world of DNA analysis gives an impressive range of insights into the details of evolutionary processes. It is applicable to the distant past – the Neanderthals are next in line for detailed analysis using DNA fragments recovered from 40,000-year-old bones. It can follow the most urgent contemporary examples of evolution, such as the changes in the viruses which cause **AIDS**, SARS or flu. Together, they add up to some of the most persuasive evidence for evolution in action that there is.

PART 3
GENES, HEALTH AND BEHAVIOUR

Genes and health

How do genes cause disease?

The ultimate aim of the Human Genome Project, and most of the numerous "post-genome" efforts now under way, is to understand the links between our genetic make-up, health and illness. Not a week seems to go by without the announcement of yet another new genetic test or breakthrough. But how has all this knowledge increased our understanding of common illnesses such as heart disease, osteoporosis or diabetes – as well as much rarer, but devastating genetic conditions such as cystic fibrosis and Huntington's disease? In this chapter, we'll look at the role of genes – and the proteins they make – both in diseases that are clearly inherited, and those that are triggered by complex interactions between genetic and other factors. We'll also cover cancer, a disease which, although it is not usually inherited, is most definitely genetic. Finally, we'll look at ageing, another biological process which, like cancer, involves the steady accumulation of genetic errors.

Genetic conditions

Genetic conditions, also called genetic disorders, are caused by changes (**mutations**) in genes that affect the way the body works or looks. Although individually quite rare – **cystic fibrosis**, for example, affects around one in every 2500 babies born – there are thousands of different genetic disorders. Taken together, this means that an estimated two to three percent of all babies are born with a genetic or chromosomal condition – approximately 13,000 births every year in the UK. And if all the genetic disorders that don't appear until adulthood are included, the proportion of affected individuals is closer to five percent. For many of these conditions, there are still no effective cures. However, the identification of many of the genes involved has improved the diagnosis, care and management of affected families, an area of medicine known as **medical genetics**.

The beginnings of medical genetics are rooted in the rediscovery of **Gregor Mendel**'s ideas at the beginning of the twentieth century. Even before then, the eighteenth-century French mathematician **Pierre Louis Maupertuis** had observed that some human traits, such as **polydactyly** (extra fingers and/or toes) and **albinism** (lack of pigment in the eyes, skin, and hair) ran in families.

Later, the English chemist **John Dalton** noticed that both the blood-clotting disorder **haemophilia** and **colour blindness** showed a particular pattern of inheritance, later called "**X-linked**", since both involve genes located on the X-chromosome. But it is generally agreed that the first recognition of a "single-gene disorder" was by UK biologist **William Bateson** and his medical colleague **Archibald Garrod**. Together, they proposed in

Pierre Louis Maupertuis

1902 that **alkaptonuria** – a condition in which a person's urine turns dark when exposed to air – was an inherited disorder involving a chemical process in the body, or, as they called it, an "inborn error of metabolism".

Garrod and Bateson described the inheritance of alkaptonuria in affected families as "recessive" – that is, the condition can appear in children born to two unaffected carrier parents. In contrast, genetic conditions described as being inherited in a "dominant" way can affect children born to families in which one parent is also affected. Such inheritance patterns are described as "Mendelian", since they follow the rules originally worked out by Mendel, from his studies of pea plants (see chapter 2). In the wake of Garrod and Bateson's insight, many other human traits were recognised as being hereditary, and by 1966 almost 1500 genetic disorders and characteristics had been identified. This prompted US doctor **Victor McKusick**, based at Johns Hopkins University, to start cataloguing all known single-gene conditions in a tome entitled *Mendelian Inheritance In Man*.

John Dalton

To keep up with the ever increasing rate of new genetic discoveries, a daily updated digital version – **OMIM** – was launched in 1987, hosted by the **US National Center for Biotechnology Information** (www.ncbi.nlm.nih.gov/entrez/query.fcgi?db=OMIM). As of October 2006, OMIM contained over 17,000 entries, which include descriptions of around 11,000 known human genes, as well as a comprehensive list of genetic conditions. It also contains a few inherited traits that aren't disorders, such as the unlikely sounding **uncombable hair syndrome**.

It's important to realize, however, that OMIM doesn't simply link individual genes to their corresponding diseases or characteristics. This is because many inherited illnesses can be caused by mutated versions of different genes – those that make proteins involved in the same biochemical pathway in the body, perhaps, or genes that encode different sub-units of the same protein. One example of this is hereditary **hearing loss**, which can be caused by mutations in any one of many different genes. The upshot of this is that two people with profound hearing loss can often have children with normal hearing. This can happen if the hearing loss is recessive (that is, only affects people who inherit two faulty copies of a gene), and the parents have mutations in two different genes. Although their children will inherit a faulty copy of both genes, they will also inherit a working copy of both, so they will have normal hearing.

Further complications arise because different mutations in the same gene can sometimes cause different disorders – for example, mutations in two genes that make Type 1 collagen (a structural protein) cause osteogenesis imperfecta, also known as "brittle bone disease". But different mutations in these **collagen genes** can cause a completely different condition, called **Ehlers Danlos syndrome**, in which affected individuals have fragile skin and unstable joints. Other faulty genes can cause similar diseases but with different degrees of severity, depending on the particular mutation – for example, Duchenne muscular dystrophy and the less severe Becker muscular dystrophy are both caused by mutations in the **dystrophin gene**.

Explaining genetic conditions

Thanks to molecular genetic techniques developed over the last thirty years or so, the genetic changes underlying many inherited conditions are now understood. Much of the early work on studying inheritance patterns in affected families helped identify these genes. Roughly speaking, those diseases described as recessive usually involve a mutation that means the gene can no longer make a working version of the protein, so individuals who inherit two such mutations (one from each unaffected parent) completely lack that protein. In contrast, dominant diseases generally involve a mutation that results in a toxic version of the protein, so that just one copy of the faulty gene (from an affected parent) "dominates", and is enough to cause the disease symptoms. A few dominant conditions are caused by the harmful effects of having a gene mutation that halves the normal amount of protein, rather than making a toxic version.

While it is true to say that all diseases that show a Mendelian inheritance pattern are "single-gene disorders", the reverse is not true – that is, many single-gene disorders aren't inherited in a straightforward Mendelian way. Some conditions vary widely in the severity of their symptoms, even between affected members of the same family. In some cases, a person may inherit the gene mutation but never display any symptoms, so that the condition appears to skip a generation – so-called **incomplete penetrance**. Often, a gene mutation can arise for the first time in the cells that make the egg or sperm, so affected children can be born to parents that have no family history of the disorder. There are other reasons why a disease may show an unusual pattern of inheritance, some of which are covered in the examples below.

We've explained the following conditions in a little more detail, some because they illustrate the range of inheritance patterns and genes involved, and some because we've mentioned them elsewhere in the book. Of course, the information provided is in no way meant to be a substitute for medical advice, and if you have any concerns about your own health or that of a family member then we recommend that you see your doctor.

Dominant conditions

As we explained in chapter 2, if one of two parents is affected by a genetic condition with a dominant inheritance pattern, every child has a one-in-two chance of being affected. So *on average* half their children will be affected and half their children will not be affected. Two examples of genetic conditions that show a dominant pattern of inheritance are Huntington's disease and neurofibromatosis.

Huntington's disease

Huntington's disease (HD) is a disorder that affects nerve cells in the brain. Mild symptoms, which include forgetfulness, clumsiness and personality changes first appear between thirty and fifty years of age. Over the next ten to twenty years, a person with HD gradually loses all control of their mental and physical abilities. In 1993 researchers discovered that HD is caused by a mutated version of a gene called **huntingtin**. The symptoms of HD appear when the abnormal huntingtin protein forms toxic "clumps" in the brain, particularly in the areas that control movement and mental abilities. There is no cure for HD at the moment, although some of the symptoms can be treated with drugs.

HD is caused by an unusual type of mutation, in which the piece of DNA code "CAG" is abnormally expanded within the huntingtin gene – a so-called "**unstable expanded repeat**" mutation, a kind of genetic stutter. In healthy people, this CAG segment is repeated 10 to 35 times, but in people with HD it is repeated 36 to more than 120 times. The abnormally expanded CAG segment leads to the production of a huntingtin protein that contains a long stretch of glutamine, one of the amino-acid building blocks that make up proteins. Unlike most amino acids, glutamine contains chemical groups which accumulate tiny electric charges and these tend to stick stretches of the protein molecules together. When the altered HD gene is passed from one generation to the next in affected families, the CAG repeat expansion often increases in size. This causes a phenomenon known as **anticipation**, in which larger repeat expansions are usually associated with an earlier onset of symptoms.

As the symptoms of HD do not usually appear until middle age, some people only discover they are at risk when one of their parents or grandparents is diagnosed. A genetic test is available to HD families that can tell people whether or not they have inherited the altered gene, but not the age at which they will start to develop symptoms. However, given the lack of effective treatments for this devastating disease, many at-risk individuals often decide not to take the test. One promising area of research into new treatments for HD aims to "silence" the faulty gene, using a new technique called RNA interference. Another involves the use of drugs that increase the cell's ability to eat up waste material in the form of **clumped protein**.

Neurofibromatosis

Neurofibromatosis (NF) is one of the most common genetic conditions. There are several different types, but about one in 2500 people has neurofibromatosis Type I (NF1). Although it can show a dominant pattern of inheritance, about half the people with NF1 are born to unaffected parents, because of a new genetic mutation in the egg or sperm. NF1 causes many benign nerve tumours, which can occur anywhere in the body, but they first appear during childhood as coffee-coloured spots on the skin. Later in life, more nerve tumours appear as lumps under the skin. It also sometimes causes mild learning difficulties.

In 1990, researchers discovered that people with NF1 have an altered version of a gene called **neurofibromin 1**, which usually helps to keep cell growth under control. Tumours appear in people with NF1 because some of their nerve cells start to grow unchecked. A genetic test may be able to detect this altered gene before symptoms appear, usually during the first year of life. As with HD, there is currently no cure for NF1, but careful monitoring can enable prompt treatment of any tumours that could otherwise cause serious complications. Research into possible new drug treatments that aim to replace the tumour-suppressing activity of the neurofibromin 1 gene is ongoing.

Recessive conditions

If two parents are both carriers of a genetic condition with a recessive inheritance pattern, there is a one-in-four chance that each child will be affected. So *on average*, one quarter of their children will be affected. There is also a one-in-two chance that each child will be an unaffected carrier, like the parents. Examples of genetic conditions that show a recessive pattern of inheritance are cystic fibrosis, sickle-cell disease, Tay-Sachs disease and haemochromatosis.

Cystic fibrosis

Cystic fibrosis (CF) is the most common serious genetic condition in the UK – around one in 22 people of Northern European descent is an unaffected carrier of the disease. CF mainly affects the lungs, where thick mucus builds up causing repeated infections and breathing difficulties, and the pancreas, where blockages cause digestive problems. Although doctors can treat the symptoms of CF, there is no cure at the moment, and the condition is life-threatening by the time the patient is thirty years old. People with CF cannot control the levels of salt and water in their body cells properly, so their body fluids are thicker than usual. They have more salt in their sweat, which doctors can measure to diagnose the condition (the "sweat test").

In 1989, researchers discovered that people with CF have no working version of a gene called **CFTR** (the cystic fibrosis transmembrane conductance regulator), which usually makes a protein that controls salt levels. Over 1500 different CFTR mutations have so far been identified in affected people, but the most common, known as DF508, results in the deletion of a single amino-acid building block in the CFTR protein. This change means that although the CFTR protein is still made by the cell, it cannot fold into the correct shape, and so is destroyed before it can get to its usual location within the cell membrane. Several research groups are aiming to develop new drugs to move the faulty CFTR protein into place, or to activate it at the membrane surface, while other drugs being developed aim to replace its function in the cell. Attempts to develop a "gene therapy" treatment for CF, in which a working copy of the gene is delivered to the cells where it is needed, have so far been confounded by the difficulties of getting enough copies of the working gene into the patients' airways, and getting the effects of the treatment to last long enough to provide any benefit. However, new ways of delivering therapeutic genes currently being developed may eventually overcome these problems (see chapter 7).

Sickle-cell disease

Sickle-cell disease (sickle-cell anaemia) affects the red blood cells, which contain a protein called **haemoglobin**. This protein is responsible for picking up oxygen from the air in the lungs and carrying it around the body. People with sickle-cell disease have an altered form of haemoglobin, which tends to join together in long strands when it is not carrying oxygen, which distort the doughnut-shaped blood cells. These "sickle" cells tend to get stuck in small blood vessels, causing pain in the chest and joints. People with sickle-cell disease often become anaemic, and may also be affected by gallstones, eye problems and strokes. Although some drugs can reduce the severity of the condition, the only cure is a bone-marrow or umbilical-cord blood transplant from a genetically matched donor.

Sickle-cell disease affects people who inherit two copies of the gene that makes the altered haemoglobin. In the UK, about one in nine people of African or Afro-Caribbean descent is an unaffected carrier of sickle-cell, although white Europeans can also be affected (we discussed the evolutionary reasons for the persistence of CF and sickle-cell in chapter 5). Screening programmes to detect the condition both in pregnancy and in newborn babies are currently being rolled out across the country. Current research is focused on drugs such as **hydroxyurea**, which appears to trigger the production of a form of haemoglobin normally only produced before birth. Boosting the levels of this foetal haemoglobin seems to prevent the red blood cells from sickling, although the long-term effects of this type of treatment are still being investigated.

Tay-Sachs disease

Tay-Sachs disease is a rare and extremely severe genetic condition that affects the brain and nerves. A baby with Tay-Sachs disease appears normal at birth, but development starts to slow down at about six months of age and the child gradually becomes blind, deaf and paralyzed. Tay-Sachs disease usually causes death before the age of five, since there is no cure for this devastating disorder, and no effective treatment. Children with Tay-Sachs disease have no working version of a gene that makes a protein called **hexosaminidase-A** (Hex-A). This enzyme normally breaks down excess fatty substances in the nerve cells. Without it, fats build up and eventually destroy the nerve cells of the affected child. This process starts in the embryo, but is not noticeable until the baby is between six and nine months old.

Tay-Sachs disease is most common in people of Central and Eastern European Jewish descent. In this community, one in 25 people is a carrier of Tay-Sachs disease, compared with one in 250 of the non-Jewish population. They are not affected by the condition, and most would never know they are carriers. But, because the disorder is so devastating, doctors offer a carrier test to high-risk communities in the UK. In the US, Rabbi Josef Ekstein founded a system called **Dor Yeshorim** (Hebrew for "generation of the righteous"), after four of his own children died of Tay-Sachs disease. It aims to identify carriers of Tay-Sachs, and other serious recessive genetic disorders, before they marry and have affected children. Teenagers are tested during large sessions in Jewish schools, and each sample is given an identifying number. When two members of the system are contemplating marriage, they contact the organization and enter their test numbers. If both are found to be carriers, they are advised to drop their plans for marriage, as each child would have a 25 percent chance of being affected.

Haemochromatosis

Haemochromatosis causes the body to absorb too much iron from the diet, and can cause a range of symptoms, including chronic fatigue, liver damage, heart problems, diabetes, joint pain and "bronzing" of the skin. If diagnosed before too much organ damage occurs, it can be very successfully treated with regular removal of blood – a procedure known as venesection therapy, or phlebotomy, which was widely used in medieval medicine but in this case has a modern application. Patients are also advised to avoid too much iron in their diets. It is usually diagnosed by measuring the amount of iron in a person's blood and confirmed with a genetic test.

Haemochromatosis is triggered by mutations in the gene that makes a protein called HFE, although unlike the other recessive disorders listed above, not everyone who inherits two faulty genes will have the disorder. Only 90 percent of men and 60–70 percent of women who inherit two mutated HFE genes (one from each parent) will actually develop haemachromatosis. The most common mutation is a single base-pair change, a "G" to an "A", which results in the substitution of the amino acid tyrosine for cysteine in the HFE protein. This in turn makes the protein unstable, so it cannot do its usual job of spanning the cell membrane and regulating iron uptake. However, there must be other genes involved in this disorder, variations in which determine whether or not symptoms develop. Strictly speaking, this means that haemochromatosis is a complex genetic disease, rather than a straightforward single-gene disorder.

X-linked conditions

Genetic conditions with an X-linked recessive inheritance pattern usually affect only the boys in a family, or, if they do affect girls, their symptoms are less severe. This is because they are caused by altered or missing genes found on the X-chromosome. Girls are born with two X-chromosomes, so the effects of a working gene on one can mask or mitigate the effects of an altered or missing gene on the other. This means that girls will normally be unaffected carriers of the condition. Boys, however, have only one X-chromosome, so if they inherit one with an altered or missing gene, they will be affected. So any son born to a carrier mother has a one-in-two chance of being affected, depending on which X-chromosome they inherit, while any daughter has a one-in-two chance of being a carrier herself. Any sons born to an affected *father* will be unaffected, since they inherit his Y, rather than his X-chromosome, while all his daughters will be carriers. Examples of X-linked recessive conditions are Duchenne muscular dystrophy, haemophilia, severe combined immune deficiency (SCID) and Fragile X syndrome. There are a few conditions that show an X-linked *dominant* pattern of inheritance – that is, a single faulty X-chromosome gene is enough to cause the disease, so equal numbers of boys and girls are affected.

Duchenne muscular dystrophy

Duchenne muscular dystrophy (DMD) is a severe muscle-wasting disorder, which affects about 1500 boys in the UK. Boys with DMD first start to show signs of muscle weakness before they are about three years old, and by twelve years they are unable to walk. By their late teens, the condition is life threatening, as the muscles used for breathing become increasingly weaker. There is no cure for DMD at the moment, although physiotherapy and exercise can help counteract the effects of the muscle weakening.

In 1986, scientists found that the muscle cells of boys with DMD are missing a large protein, called **dystrophin**, which provides them with vital structural support. Without it, the cell membrane becomes "leaky", allowing substances outside the cell to enter until it eventually "explodes" and dies. In the less severe **Becker muscular dystrophy**, patients still make the dystrophin protein, but it doesn't work properly. In about a third of cases, DMD appears "out of the blue", following a new mutation in the unfertilized egg. In others, the disorder is inherited from a carrier mother, who is usually unaffected herself.

Efforts to develop a gene-therapy treatment for DMD have been hampered by the unusually large size of the dystrophin gene, which is too big to deliver using standard methods. But recent research using a "mini-dystrophin" gene holds promise, as do possible new treatments based on boosting the level of other proteins in the cell to replace the missing dystrophin. Work on stem-cell-based therapies for DMD is also ongoing.

Haemophilia

Haemophilia is a blood-clotting disorder that mainly affects boys. People with haemophilia bruise easily and bleed for longer if they injure themselves, but the main problem is bleeding inside the body, especially around joints. There are two main types of the disorder, which together affect about 6500 people in the UK. Both are caused by gene mutations that affect blood clotting factors. In some cases, the disorder is caused by a new gene mutation in the unfertilized egg. In others, the affected boy inherits haemophilia from his carrier mother, about a third of whom have mild symptoms themselves.

Since the 1970s, doctors have successfully treated people with hae-mophilia by injecting them with clotting factors purified from donated blood. Unfortunately, many patients have caught viral diseases after receiving clotting factors made from infected blood. In the UK, over 1200 men were infected with HIV during the early 1980s in this way. To protect against this risk, and also the theoretical risk of **Creutzfeldt-Jakob disease** (CJD), clotting factors produced by genetically engineered bacteria can be used instead. Since 1998, all UK children under sixteen have been treated with these "recombinant" clotting factors, and they are now available to everyone with the condition.

Severe combined immunodeficiency

Severe combined immunodeficiency (SCID) is a very rare and serious condition that affects the immune system. X-linked SCID, the most common type, usually affects only boys. Without treatment, even a common cold can be life threatening for a child with SCID, so until they can be treated, children diagnosed with the disorder must be looked after in a sterile environment, or "bubble". In some cases, the disease is caused by a genetic change in the unfertilized egg, while in others, the affected boy inherits SCID from his unaffected carrier mother.

X-SCID is caused by mutations in the IL2RG gene, which makes part of a receptor protein that directs the growth and activation of immune system cells. Affected boys cannot make a type of white blood cell called T lymphocytes, while another type, the B lymphocytes, cannot make essential antibodies for fighting infections. The only effective treatment for SCID is a bone-marrow transplant from a genetically matched donor, to replace the white blood cells. Some boys with X-SCID have been successfully treated with gene therapy, although in a French trial three of the participants subsequently developed leukaemia, one of whom died. It seems that the virus used to deliver the therapeutic gene activated a cancer-causing gene, a side effect that does not seem to have arisen in trials taking place in other centres around the world. (See chapter 7 for more on gene-therapy research).

Fragile-X syndrome

Fragile-X syndrome is the most common cause of inherited mental impairment, affecting about one in 2500 boys and about one in 5000 girls. As well as learning difficulties, people with Fragile-X may have other symptoms, including anxiety and hyperactivity. Boys are usually more severely affected than girls. While there is no cure for Fragile-X syndrome diagnosis is important, as special education and therapy can help affected children.

In 1991, researchers found that people with Fragile-X have an altered version of a gene called FMR1, which makes a brain protein. Like Huntington's disease, the mutation is an unstable expanded repeat, this time involving the DNA code "CGG". Individuals with fewer than 60 CGG repeats are unaffected, while those with 60–200 CGG repeats have a "pre-mutation" that can expand in future generations. Individuals with over 200 repeats have a full mutation, which causes symptoms of the condition. The full mutation causes the gene to "shut down", via chemical tags called methyl groups that attach to part of the FMR-1 gene. When the gene is turned off, the individual does not make a protein called **Fragile-X mental retardation protein** (FMRP1). Since the FMR-1 gene is on the X-chromosome, girls will normally have a working copy of the gene that partially makes up for the mutated gene, so they are usually less severely affected than boys.

It seems that the normal role of FMRP1 is to help the brain cells that underlie learning and memory communicate with each other, by controlling the activity of other genes. Current research is focused on finding ways to restore these connections.

Other unusual inheritance patterns

As described above, new mutations, incomplete penetrance, variability and anticipation can all result in an unusual inheritance pattern that may be difficult to interpret. Another example of "non-Mendelian" inheritance results from genetic imprinting, which refers to genes whose activity is determined by which parent they were inherited from. Imprinting involves the "tagging" of certain genes in mammals (including humans) during sperm and egg production to mark them as either coming from the mother or the father. The tagged copy is then permanently switched off in some or all of the tissues where the gene is used, so that the cell only uses *either* the copy from the mother *or* that from the father to make proteins. Examples of conditions that involve imprinted genes are Angelman syndrome, Prader-Willi syndrome and Beckwith-Wiedemann syndrome.

Some unusual inheritance patterns can be accounted for by **mosaicism**, in which the disease-causing mutation arises *after* fertilization. If a gene acquires a mutation during early embryo development, then it will only affect those tissues and organs produced by the descendants of the cell in which the mutation arose. This results in an individual who has the mutation in some parts of their body, but not others. Mosaicism is distinct from **chimerism**, although this too results in an individual with genetically different cells in different parts of their body. But chimeras are the result of two embryos fusing very early in development (the opposite of twinning), and so the tissues that form tend to show many distinct genetic differences, rather than just one mutation as is the case with mosaics. Finally, mutations in the DNA of the mitochondria, the cells' powerhouses, result in a distinct maternal inheritance pattern (see box on p.128).

Treating genetic conditions

As shown by the examples above, while there are few cures for genetic conditions, many can be effectively treated or managed. Although gene therapy – replacing a missing or faulty gene with a working copy – is sometimes hailed as the ultimate cure for many inherited diseases, it is likely that it will only ever be suitable for a small number of diseases that affect just one or two tissues (not counting the promising gene-based therapies aimed at treating cancer). In the meantime, there are a variety of other approaches for treating genetic conditions that are based on more conventional medicine. Some disorders can be treated with a synthetic version of the missing protein: blood-clotting factors for haemophilia, for example. And for conditions that affect the blood, such as sickle-cell disease or the severe immune disorder **X-SCID**, replacing all the affected cells via a bone-marrow transplant can sometimes result in a complete cure.

Some conditions can be helped with organ transplants; for example, lung transplants can help people with the lung condition **cystic fibrosis** (CF). However, finding a donor organ with the same tissue type can be difficult, and patients still need to take drugs to stop their body rejecting the transplant. In the meantime, CF patients rely on a cocktail of enzymes, to help them absorb nutrients, daily physiotherapy and strong antibiotics to combat lung infections. Early diagnosis is the key to the effective treatment of CF, and any other genetic condition that can be ameliorated by

Mitochondrial disorders

Most of a human cell's genetic information (about 99.9995 percent) is contained within the nucleus, bound up in its 46 chromosomes. But the remaining 0.0005 percent is contained within the mitochondria – tiny sausage-shaped structures that supply the cell with energy. Depending on its role in the body, and how much energy it requires, one cell can contain hundreds or thousands of mitochondria. The DNA in the mitochondria encodes just 37 genes, of which 13 are instructions for making parts of proteins needed to release energy from glucose and other fuel molecules. However, this complex set of chemical reactions involves a total of about one hundred different protein subunits, most of which are encoded by genes in the cell nucleus. So to make energy (in the form of a chemical called **adenosine triphosphate**), the cell needs mitochondria but equally, mitochondria need the cell.

Anywhere there's a gene, there's the possibility of mutations – and mitochondrial genes are no exception. In fact, mitochondrial DNA seems to be more error-prone than DNA in the nucleus, probably because it is copied less carefully and much more often. But most of these errors will have no effect, since a cell has so many mitochondria that a single faulty one will not cause any problems. The one exception to this is in the cells destined to form egg cells, since if they are fertilized, the mitochondrial mutation will be passed on to all the cells of the developing embryo. All of a sperm's mitochondria are located in its tail, which falls off after fertilization, so the mitochondria in your body are all genetic copies of your mother's. Because of this, disorders caused by mutations in the mitochondrial genes show a distinctive inheritance pattern, in which the condition is only passed on by mothers, although equal numbers of girls and boys are affected. Examples of mitochondrial disorders include a disease that causes sudden loss of vision, called Leber hereditary optic neuropathy, and several different forms of inherited muscle weakness. The full mitochondrial DNA sequence and the mutations and variations identified so far are all catalogued in the MITOMAP database (www.mitomap.org).

diet, medicines or educational support. The best hope for many genetic disorders is more research into their underlying causes. Identifying the gene involved is just the starting point for a better understanding of how the symptoms arise, and how they might be treated.

Chromosomal conditions

As well as the disorders listed above, which are all caused by alterations in a single gene, babies can be born with conditions that involve alterations of **chromosomes**, whole bundles of genes. When cells divide to make egg and sperm, each new cell usually ends up with 23 chromosomes, one from each pair. Sometimes, however, part or all of a chromosome is missing, rearranged or duplicated. Most fertilized eggs that have more or fewer than 46 chromosomes will not survive. In some cases, however, the embryo will survive, but the chromosomal alterations may affect the way it grows and develops. The best-known chromosomal condition is Down syndrome, which is usually caused by an entire extra copy of chromosome 21, the smallest human chromosome. This is responsible for the features of Down syndrome, including delayed development, some learning difficulties and possible heart problems. The chances of a couple having a baby with Down syndrome depend on the mother's age. For women under the age of 25, the chances are about one in 1400, rising to one in a 100 by age 40.

Down syndrome is an example of **trisomy** – having three copies of a particular chromosome. Two others are Edward syndrome (three copies of chromosome 18) and Patau syndrome, or trisomy 13. Most other trisomies result in early pregnancy loss, particularly chromosome 16, which is a common cause of miscarriages that occur during the first twelve weeks. Trisomies are usually caused by the failure of a pair of chromosomes to separate in the dividing cells that produce egg and sperm (see chapter 1). Other chromosome abnormalities include **monosomy** – having just one copy of a chromosome, although the vast majority of such embryos do not survive. An exception is Turner syndrome, which affects women who are born with one X-chromosome rather than two.

Both trisomies and monosomies are "one-off" occurrences that do not run in families. However, other types of chromosome alterations can sometimes be present in healthy carriers, who then have a higher chance of giving birth to an affected child. Such alterations include **translocations**, in which parts of two different chromosomes have switched places, and **inversions**, which, as the name suggests, involve part of a chromosome that is flipped around in its usual location. A few cases (3–5 percent) of Down syndrome are caused by translocations involving chromosome 21 and another chromosome, usually 14. These individuals are clinically indistinguishable from those with trisomy 21, but about a third of the parents of translocation cases have a relatively high chance of having another affected child. However, the majority of translocations - as well as inversions and other chromosome alterations such as deletions (small bits of chromosome missing) and insertions (small bits of chromosome inserted) – occur for the first time in an affected child, and so are not associated with a high recurrence risk.

Genes and common disease

Encouraged by their success in identifying the gene mutations that cause genetic conditions, back in the mid-1990s many scientists had high hopes of unravelling the genetic influences on common diseases such as heart disease, asthma and diabetes. But ten years on, and even with the entire human genome sequence at our disposal, it's fair to say that these so-called "complex" disorders remain just that. Genes do affect our chances of developing many common illnesses, as numerous studies of twins and adoptees have shown (see box on p.134), but it's a bit like betting on a horse race – the horse, rider, course and weather can all affect the outcome in a way that is hard to predict. For example, you may have a gene variant that increases your risk of Type 2 diabetes, but your diet, how much you exercise and whether or not you smoke also affect your chances, as well as other genetic variations that may either increase or decrease your risk.

In addition, since genetic predispositions to disease are dependent on many other factors, they won't always be apparent. In his book *Your Genes Unzipped*, UK twin researcher **Tim Spector** uses the analogy of a non-swimmer drowning in a swimming pool: if we think of the person's height as the genetic risk factor, and the depth of the water as the environmental one, then at moderate depths genes for shorter height increase the risk of drowning. But in very shallow, or very deep water, the height genes become irrelevant. The same principle applies to a condition like severe obesity – if everyone ate far more than they needed, then obesity would afflict us all. At the other extreme, it obviously doesn't affect people who do not have enough to eat. The reality in most countries is somewhere in between – so at least some of the variation in populations is down to genes. In summary, many genetic and non-genetic factors (sometimes called "environmental" factors) affect our health, but at the moment we're far from knowing what they all are, or how they interact with each other.

Most of the gene variants involved in common disease identified so far account for relatively small increases and decreases in risk. Even where such variants are identified in one population, they often fail to show an association with affected people in a different population. There have, however, been some notable successes in the identification of rare forms of common conditions. For example, a few severely obese people have mutations in the gene that makes **leptin**, a hormone involved in controlling appetite and storing fat. If identified, children with these mutations can be very successfully treated with leptin injections, which over time can reduce their weight to near-normal levels.

It was hoped that variations in such genes, identified through their involvement in rare versions of common diseases would account for the majority of the genetic risk in the rest of the population. Unfortunately, this has not often proved to be the case. Instead, geneticists are still spending long hours trawling for elusive gene variants, each of which has only a slight effect on a person's risk. Below are examples of the progress made so far in understanding the genetic influences on four common diseases.

Alzheimer's disease

Alzheimer's disease (AD) is the most common cause of dementia, a slow decline in mental ability. It affects one in twenty people over 65, and more than one in ten of those over 85. AD is caused by the gradual death of certain brain cells, especially in the areas involved in memory. Most people with AD have not inherited it, but in a few cases (less than one in one hundred), the disease is caused by a gene mutation inherited from an affected parent. In this type of Alzheimer's the symptoms tend to appear earlier than usual. Researchers have also found a version of another gene, called **APOE4**, which is associated with an increased risk of AD in non-familial cases. This gene alone seems to account for about fifty percent of the genetic susceptibility to late-onset AD – people who inherit two copies of APOE4 (one from each parent) have around fourteen times the risk of people who inherit no copies of this variant. But, as with other so-called "susceptibility genes", not everyone who inherits a double dose of APOE4 will develop AD, so other genes are clearly involved.

Heart disease

In people with heart disease, the coronary arteries become "furred up" with fatty deposits, so that not enough oxygen reaches the heart. A person may suffer a heart attack if the coronary arteries become completely blocked. Heart disease has many causes, including a high level of cholesterol in the blood, smoking, and high levels of substances that usually help the blood to clot. But in some families, heart disease is caused by gene mutations that affect the levels of cholesterol in the blood. People with familial hypercholesterolaemia are born with twice the normal level of cholesterol in their blood. Other people have gene mutations that increase blood pressure, or the tendency of the blood to form clots.

The APOE4 gene variant linked to Alzheimer's disease is also a susceptibility gene for **atherosclerosis**, to give artery furring its proper name, which in turn can increase the risk of heart attacks and strokes. This makes biological sense, since the APOE4 gene makes a protein called **apolipoprotein E**, which plays an important role in removing excess cholesterol from the blood and carrying it to the liver for processing.

Allergies

Allergies, particularly **asthma**, have increased dramatically during the last couple of decades – a rise that is too recent to be explained by genetic changes. But genes are definitely involved, since allergies tend to run in families, and those affected by asthma are also more likely to suffer from **hayfever** or **eczema**. For most people, the risk of developing asthma is around one in twenty, but if one of your parents has asthma, your risk rises to one in four. Allergies can have many different causes, for example, pollen, house dust mites or animal fur. Some scientists think that our use of antibiotics and antiseptics has produced an increasingly "germ-free" environment, which makes our immune systems overreact to such seemingly harmless invaders. But whatever the trigger, asthma may still only affect those who are already genetically susceptible. It's estimated that around ten different genes influence the risk of asthma, of which about half have already been identified.

The efforts of eczema researchers have also recently been rewarded, with the identification of gene mutations that affect a protein called **filaggrin**, which normally helps form the skin's tough, protective barrier that keeps infections out, whilst keeping water in. It seems that filaggrin mutations could be involved in triggering eczema in up to half of the people who have the condition, as well as in some people who inherit both asthma and eczema.

Diabetes

There are two main types of diabetes, a common condition caused by resistance to, or lack of **insulin** – the hormone that controls blood-sugar levels. Type 1 diabetes usually first appears in childhood, and can only be treated with regular injections of insulin. In contrast, Type 2 diabetes usually only affects people over the age of forty. It can be treated by changes in diet and medication, although some patients eventually also require insulin. In Type 1 diabetes, the cells in the pancreas that make insulin are destroyed by the body's own immune system. In Type 2 diabetes, either the pancreas does not make enough insulin or the body becomes resistant to its effects. Both types run in some families, but genetic factors play a greater role in the onset of Type 1, whilst for Type 2, other factors such as diet seem to be more important.

Much of the genetic risk for Type 1 diabetes is accounted for by versions of two "HLA" genes that make proteins that label cells as "self", to prevent their destruction by the immune system. In people with the condition, this labelling system goes awry, leading to loss of the pancreas cells. But not everyone who has these two variants develops the condition, so there must be other genes and non-genetic factors involved. For Type 2 diabetes, several gene mutations have been linked to rare, childhood-onset forms of the condition, known as maturity-onset diabetes of the young (MODY). However, only one major gene variant that significantly affects the common, adult form of the disease has been identified so far – a version of the **TCF7L2 gene**, which makes a protein involved in controlling blood sugar. But as with other susceptibility genes, although this variant is present in many people affected by Type 2 diabetes, not everyone who inherits it will develop the condition.

Why look for genetic influences on disease?

Despite the difficulties, searching for genes that may influence our risk of getting ill is currently an intensely busy field of research. But supposing these efforts do meet with success – what next? At the moment it seems unlikely, despite what is often suggested, that this endeavour will pave the way for "tailored risk assessment" – the idea that doctors will one day be able to take a look at your genetic information and then advise you to, say, eat more fruit and fewer cream cakes to cut your hereditary risk of heart disease. (This hasn't prevented an industry keen to provide such **nutrigenomic tests** from springing up – see chapter 14). This is because genetic information of this sort is unlikely to add much to that which can be gleaned from your family health history – if heart disease runs in your family, you are likely to be at increased risk too. Since genes are only *part* of this risk, a healthy diet and active lifestyle can drastically cut your chances of being affected – but this is general good advice that applies to everyone, not just people carrying a particular gene variant.

However, the identification of genes involved in complex diseases *will* shed light on their underlying biology, which in turn might help patients. If, for example, scientists could work out exactly which brain proteins are involved in schizophrenia, then it could lead to new treatments and more effective use of existing treatments, by matching different drugs to people affected by different sub-types of the condition. As with identify-

How is it done?

So how are scientists seeking the elusive genetic variations that affect our health? The first step is to figure out which conditions are influenced by genes, by studying affected families, adoptees, or twins. A family may share its genes, but they also share many other things, including beliefs, diet and hobbies – supporting Arsenal "runs in families", but no one has (yet) suggested it might be genetic. To overcome the limitations of looking at families, many studies have compared identical twins, who have identical genes and similar backgrounds, with non-identical twins – who also share similar backgrounds, but are no more genetically alike than ordinary siblings. In this way, they can estimate how much of the variation seen for a particular characteristic is accounted for by genes – a measure of "heritability". A trait that is completely controlled by genes, like eye colour, has a heritability of one hundred percent. At the other extreme, a characteristic that has nothing to do with genes, such as a scar, will have a heritability of zero. Such studies show, for example, that Alzheimer's disease (not the rare familial kind) has a heritability of sixty percent – in other words, genes contribute slightly more to the risk of this condition than non-genetic factors. However, twin studies have been criticized for not being representative of the general population, since twins share a womb, as well as their genes. For this reason, scientists have also looked at adoptees, and asked if they are affected by diseases that run in their biological, rather than their adoptive families. But as with twins, the numbers of such people are quite small, and the proportion of those that have good health information about their biological families is smaller still.

Assuming there's good evidence that something is inherited, the next stage is research: large-scale studies looking for links between particular genetic variations and the disease. Large numbers are needed because there may be more than a hundred different genes involved in a common condition such as heart disease, each one contributing a tiny effect. These studies can involve lots of affected families, trying to track particular variants and measure whether they are inherited with the disease more often than expected by chance. Or, they may simply involve large numbers of unrelated, affected cases and unaffected people as controls. Both types of study require some serious statistical analysis, carried out with the aid of computers, to see any significant effect. Things have speeded up in recent years with the production of a catalogue of human genetic variations, the **HapMap** (www.hapmap.org – see p.111). Combined with machines that can look at many of these variations at once, such "whole genome" scans are becoming an increasingly efficient way to identify regions of DNA that contain genes involved in a particular condition. Another way to get at disease genes is to avoid human subjects altogether, and study animals bred to have symptoms of the condition. This approach obviously works better for a physical condition like diabetes, rather than a mental illness such as schizophrenia, which is difficult to diagnose in people – never mind mice. Even once a risk variant has been pinpointed, however, much more research is needed to find out whether it is genuinely associated with a disease, and what biological effect it might have.

ing the faulty genes that underlie simple genetic conditions, homing in on the genes (and the proteins they make) involved in common, complex disorders is just the first step down the road to improved health care. Such discoveries, if they are made, could eventually have a profound effect on 21st-century medicine – a topic we cover in more depth in chapter 14.

Understanding complex diseases will increasingly involve the study of gene activity – where and when particular genes are switched on in the body, and which factors can affect these switches – rather than the genes themselves. This area of research, called **epigenetics** (see chapter 4), helps explain why identical twins are not always identical, despite having exactly the same genes. A recent study showed that the "epigenetic profiles" of identical twins show increasing variation as the twins age, suggesting that the individuals have different patterns of gene activity, presumably caused by different environmental effects. This in turn could affect their risks of developing certain diseases.

It used to be thought that these epigenetic switches were all "reset" – a bit like a computer being returned to its factory settings – when sperm and eggs are made. But other recent evidence suggests that like gene mutations, some epigenetic changes may be passed from one generation to the next. It was already known that a pregnant woman's diet can affect the health of her future child. Now, recent research suggests that a *father's* diet and lifestyle can also affect his children's gene activity. What's more, studies carried out by researchers in the UK and Sweden suggest that such "transgenerational effects" may even apply to grandparents. A famine, for example, may have lasting effects on gene switches which regulate metabolism.

Cancer and genes

Cancer is usually the result of genetic damage acquired during a lifetime, which is why most cases occur in people over sixty. Your body is made up of 100 million, million cells. Cancer can start when just one of them begins to grow in an uncontrolled way, forming a **tumour**. This amounts to a cell reverting to the way it behaved before it was part of a multicellular organism – because evolution depended on differential reproduction, every cell's built-in ambition is still to become two cells. Our bodies have developed elaborate systems to regulate this unruly behaviour. But they sometimes break down.

Benign tumours are localized growths – they only cause problems if they put pressure on nearby tissues, such as the brain. Much more serious are **malignant tumours**, which invade the surrounding body tissues. Some malignant tumours also spread throughout the body via the bloodstream: a process called **metastasis**. You have two hundred different types of cell in your body, which usually all live, grow and multiply in harmony – you always have just the right number of liver cells and white blood cells, for example, because there are many different proteins that control how much and how often your cells divide. Some act as brakes on cell growth, while others are accelerators. If any of these proteins are faulty or missing, because a mutation arises in the gene that codes for them, then a cell may start to grow and multiply too much.

Most genetic errors arise when a cell divides to make two new cells, which involves copying all 6000 million DNA base-pairs of its genetic code (on two sets of chromosomes). Inevitably, mistakes are sometimes made, causing mutations. Most of these are corrected immediately, but a few manage to escape unnoticed. But even once a mutation has occurred in a gene that makes a **growth-control protein**, a healthy cell does not turn into a cancer cell overnight. Its behaviour gradually changes, a result of damage to between three and seven of the hundreds of genes that control cell growth, division and life span. Over time, these genetic changes mean that the cell and its descendants may eventually become immortal, escape destruction by the body's defences, develop their own blood supply and invade the rest of body: the hallmarks of a late-stage cancer.

It might seem very unlikely that a single cell could acquire mutations in all the genes necessary to make it cancerous - and it would be, if it were simply a matter of chance. However, one of the features of a cancer cell is that it gradually loses the checking mechanisms that normally ensure genes are accurately copied each time a cell divides. The result is chaos: genetic material may be lost, rearranged or copied many times and the genes are more likely to acquire further mutations. For example, an important protein called **p53** normally checks for gene damage in healthy cells, and kills them if the damage is too great to repair. However, cancer cells often have a faulty p53 protein, which allows cancer cells to survive despite having dangerously garbled genetic material. Also, if a cell acquires a gene mutation that makes it multiply when it should not, or helps it survive when other cells die, it has an advantage over the others. Eventually, the abnormal cells acquire mutations in more genes, causing uncontrolled growth. These abnormal cells then have a competitive advantage over normal cells, and their descendants are more likely to acquire further mutations. This process of stepwise change sometimes takes many years. Most men over fifty have some of the early changes which lead to **prostate cancer**, for example, but the majority of them will die from some other cause before a prostate malignancy develops.

One of the mutations that leads to malignancy causes "immortality", allowing malignant cells to escape destruction. Every time a normal cell divides, the ends of its chromosomes, called **telomeres**, become shorter. Once their telomeres reach a certain length, the cell dies and is replaced. Cancer cells cheat this system – they retain their long chromosomes by continually adding bits back on, allowing them to become immortal. Cells from Henrietta Lacks (HeLa cells), an American woman who was diagnosed with cervical cancer in 1951, are still growing in culture. The cells were taken and grown by George Gey, head of tissue culture at Johns Hopkins University, who had been looking for a sample of cancer cells to grow in the laboratory for his research. Sadly, the cells grew as quickly inside Henrietta's body as outside of it, and she died of her illness a few months after the sample was taken. Although many human cell lines have since been cultured, HeLa cells remain one of the most robust, and have proved vital in many areas of medical research – not just cancer. You can read a fascinating account of how Henrietta's family discovered her cells were still living, many years after her death, in a 2000 article published by the Johns Hopkins magazine: www.jhu.edu/~jhumag /0400web/01

Another key feature of cancer cells is the ability to break free from the tissue of which they are a part. Most normal cells stay put, stuck to each other and their surroundings. Unless they are attached to something, they cannot grow and multiply. If they become detached from their neighbours, they "commit suicide", by a process known as **apoptosis**. But in cancer cells the normal self-destruct instructions do not work, and they can grow and multiply without being attached to anything. This allows them to invade the rest of the body, travelling via the bloodstream to start more tumours elsewhere (metastasis), the final stage in a cancer.

Cutting the risk of cancer

So cancer, therefore, is abnormal cell growth, which is the result of mutations in certain crucial genes. There are several ways in which you can limit this gene damage: by avoiding cigarette smoke and other harmful chemicals, eating a healthy diet with plenty of fruit and vegetables and limiting your exposure to the ultraviolet radiation that causes sunburn.

It's estimated that a third of all cancers in the UK are linked to poor diet, while a further third are triggered by smoking. Some types of food increase your risk, while others seem to have a protective effect. For example, a diet low in saturated fat, alcohol and red meat and high in fibre, fruit and vegetables can reduce your risk of some cancers, such as breast and colon cancer. Fruit and vegetables contain natural chemicals called **antioxidants** that mop up "**free radicals**" – highly reactive chemicals that can damage genes. Some viruses also trigger cancer; for example, cervical cancer is linked to **human papillomavirus** (HPV), which causes genital warts. For this reason, there have been motions for girls in the UK and elsewhere to receive a HPV vaccine before they become sexually active and risk exposure to the virus.

Usually cancer is not inherited, even in families with several people affected, since it is a very common disease. However, some people *are* born with an increased risk of cancer, because they inherit a mutation in a gene important for cell growth or for repairing damaged DNA. This means that all the cells in their body have already taken one step down the multistep pathway that turns a normal cell into a cancerous one. Scientists have already identified some of the genes that predispose to breast cancer and colon cancer, but there are likely to be several others.

Hereditary breast cancer

Breast cancer is common – it affects one in eleven women at some time in their life – so in most cases, a family history of the illness does not mean it is inherited. But five to ten percent of affected women do have an inherited form of the disease, usually with earlier onset. Scientists have identified two genes that are altered: **BRCA1** and **BRCA2** (see p.70). These genes normally make proteins that repair DNA damage, preventing abnormal cell growth. A woman who inherits a mutated version of either of these genes has a high risk of developing breast cancer. Anyone with a strong family history of breast and/or ovarian cancer, particularly if they have relatives who were affected when young, is advised to speak to their doctor about concerns over their own health.

Colon cancer

Most colon cancer is not inherited, but is linked to less healthy low-fibre, high-fat diets. However, five to ten percent of affected people do have an inherited form of the disease. Scientists have identified several genes involved in hereditary colon cancer. These genes normally make proteins that repair damage to other genes, preventing abnormal cell growth. A person who inherits a mutated form of any of these genes has a high risk of developing colon cancer, regardless of their diet and lifestyle. As with hereditary breast cancer, anyone concerned about a family history of colon cancer should speak to their doctor.

Cancer and genetic research

It's been over thirty years since US President **Richard Nixon** declared his famous "War on Cancer" in 1971. Shouldn't scientists have won by now? In fact, there have been some major advances in the prevention, early detection and treatment of cancer, with the result that many people who would have died from the illness thirty years ago now survive. But in reality, what we call cancer is actually over two hundred different diseases, since there is at least one type for every different tissue in the body. Each one involves different proteins and responds to different drugs. Even cancers that affect the same tissue differ widely in how aggressive or treatable they are. Despite these major challenges, many doctors hope that cancer will eventually become a lifelong condition that can be controlled – provided we can afford the drugs – rather than a fatal disease.

Genetics has a major role to play in this research, since the key to understanding the biology of different types of cancer is working out which growth-control genes are at fault – both those that trigger the cancer, and those that are involved during the later stages of malignancy. This work is helping scientists to develop new tests that can pick up cancer in its earliest stages, and new treatments that target its weak spots. The **Cancer Genome Project**, based at the **Wellcome Trust Sanger Institute** in Cambridge (www.sanger.ac.uk/genetics/CGP), aims to catalogue all the genetic mutations that occur in different types of cancer cell. The US Cancer Genome Atlas (cancergenome.nih.gov) is aiming to do much the same, but on a larger scale. The scientists hope that by taking this systematic approach, they will be able to find out precisely how aggressive a particular cancer is likely to be, and which drugs will be most effective against it.

Ageing

We're living longer and healthier lives than ever before – as recently as a century ago, the average lifespan in the UK was only 49 years for men, and 52 for women. Many children died before the age of five. Today, thanks to steady improvements in diet, health care and living conditions, the average lifespan has risen to 74 for men and 79 for women. Although some of the shift in the average is due to fewer deaths in infancy, many more people are also living to a ripe old age. The UK Government's Actuaries Department predicts that by 2031, the average lifespan will be 79 for men and 83 for women. But despite this astonishing extension of our lives compared to that of our great, great grandparents, ageing – and ways to extend a healthy life still further – is a very active area of research. So what makes us age, and can we do anything about it?

Your life is not programmed to end, you simply wear out. The ageing process – thinner, drier skin, weaker bones and muscles, memory loss and a less efficient immune system – all comes down to the steady accumulation of genetic damage. Apart from brain cells, most of the cells that make up your body are constantly replaced, as existing cells multiply to make new ones. But every time a cell divides, the ends of its chromosomes, called **telomeres**, become shorter. Once the telomeres reach a certain length, the cell stops dividing and eventually dies. The only healthy cells to escape this fate are those that divide to make eggs and sperm, in which a substance called **telomerase** builds the telomeres up again. Scientists have made normal cells in a test tube multiply indefinitely by adding telomerase. Some have claimed that telomerase could be used as an elixir of life, but this is unlikely. Apart from germ cells, the only other cells that can multiply indefinitely are **cancer cells**. Allowing old, damaged cells to carry on growing and dividing forever would probably have disastrous results.

As more cells are lost or damaged, you start to show signs of ageing. Cells are damaged by the harmful molecules that are continually bombarding your body, particularly glucose and "free radicals", a by-product of energy production. Glucose, and other sugars such as fructose, attack proteins in particular, a process called **glycation**. Damaged proteins are less effective, which compromises the ability of the cells to work properly. Free radicals damage proteins too, but they also attack DNA. In his book *Power, Sex And Suicide*, science writer and biologist Nick Lane argues that this is the price we pay for hosting the mitochondria that allow us the efficient energy release from oxygen, and probably enabled multicellular organisms to evolve in the first place.

When you are young, your body is able to repair most of this damage. But as you get older, the repair process is less efficient. Intriguingly, mice which were fed thirty percent fewer calories than they would usually consume live longer, perhaps because they are burning less food, and so producing fewer free radicals. Another theory is that their bodies are adapting to the lack of nutrients by slowing down ageing, so they can reproduce when conditions improve. The key to this adaptive response could be a set of genes known as **sirtuins**, which help cells to withstand stress. Sirtuins are currently the focus of much research, as when manipulated they can extend the lifespan of nematode worms and flies. Other genes thought to play a role in ageing include those that are directly involved in fat metabolism, those that protect cells from genetic damage, and those that repair DNA. Most ageing research is aimed at tackling the diseases that appear in old age, so that people can lead healthier, better quality lives, rather than greatly extended ones. We may just have to accept we've evolved to live long enough to reproduce and care for our offspring. Once we've successfully passed on our genes, there's no reason for natural selection to favour mutations that affect our subsequent survival. But this inconvenient biological reality is unlikely to stop large amounts of money being poured into ageing research, much of it privately funded – for example, by **Aubrey de Grey**'s SENS (Strategies for Engineered Negligible Senescence) foundation. De Grey, a Cambridge scientist and engineer, co-founded the Methuselah Mouse Prize, which provides cash incentives to researchers who find ways of extending the lifespan of mice to unprecedented lengths. (See www.sens.org)

Genetic medicine

Genetic information provides the code for all the proteins your body needs to grow, develop and work properly. As we have seen, if this information is corrupted then it can have serious health consequences. But now that scientists have begun to understand this code, and the ways in which it is interpreted by our cells, surely there must be new treatments for diseases that have a genetic basis?

The answer, unfortunately, is still "no" for most genetic conditions, although genetic testing is now a crucial part of diagnosis. And couples at risk of having a child affected by a serious genetic disorder may opt for genetic tests during pregnancy (prenatal tests) – or, in some cases, embryo tests combined with **in vitro fertilization** (IVF). This last development has triggered media hysteria over so-called "designer babies", despite its limited use at present. Both prenatal and embryo testing, along with screening programmes to detect serious genetic disorders, have also sparked lengthy debates on whether modern genetic medicine represents a resurgence of eugenics.

For a small number of patients affected by serious genetic conditions, gene therapy has provided an effective new treatment – although this approach is still very much in its infancy, and not without risks. In fact, the most promising application of gene therapy at the moment is in the development of new treatments for cancer. For other common diseases such as diabetes and heart disease, genetics is just one of many areas of research that is shedding light on their underlying biology. Many scientists hope that identifying the different genes involved in these illnesses will allow them to target drug treatments more accurately – see chapter 13 for a detailed discussion of this "**pharmacogenetic**" approach. But this chapter will focus on the possibilities and limitations of current, rather than future "genetic medicine", as well as the concerns that it raises.

Genetic tests and genetic counselling

In many ways, genetic tests are just another sort of medical test, used by doctors to help them make a diagnosis of a genetic condition. This process also involves carrying out other tests, to look for specific signs and symptoms, and taking a detailed family history. As a result, other family members at risk of the condition may be contacted, with the patient's consent. So, *unlike* other sorts of test, a genetic test can sometimes have a "ripple effect" that spreads throughout the family. In the UK, genetic tests are usually carried out at specialist genetics centres, following a referral from the patient's family doctor. Since the 1980s, there has been a rapid increase in the number of different DNA-based tests available, and there are now around 25 such centres throughout Britain (for details see the **British Society of Human Genetics** website, www.bshg.org.uk). The North West Regional Genetic Service, for example, which covers Manchester in England and the surrounding towns and cities, saw 8,300 patients in 2005, and carried out 36,000 DNA tests. As well as performing genetic tests, specialist centres aim to provide emotional and medical support for affected individuals. Patients and their families are offered genetic counselling, which is not the same as counselling in the usual "therapy" sense of the word. Genetic counsellors talk to patients and their families about the implications of a genetic test result, and can provide information on risk to other family members. This information varies, depending on who is being tested, and for what reason, but some typical cases might be:

▶ A child with delayed development and behaviour problems is referred with a suspected diagnosis of Fragile-X syndrome, which can be confirmed or ruled out with a genetic test.

▶ A man has just been diagnosed with a hereditary form of bowel cancer at the age of 25, and his two sisters want to know if they have inherited the gene mutation that triggered the disease.

▶ A woman has a nephew who has Duchenne muscular dystrophy, and wants to know if she too is at risk of having an affected child.

▶ A family have a child with Tay-Sachs disease. The woman is pregnant again, and the couple want to know if the baby will be affected by the same condition.

As these four examples show, genetic tests can be done to confirm a diagnosis; to predict who might be at risk of developing a condition; to find out if someone is an unaffected "carrier" of a genetic condition; and to find out if a condition affects an unborn child. Genetic counsellors explain the results of these tests, and help families decide what course of action to take, if any. Genetic counselling aims to be totally non-directive, which means that no one is "directed" or told what decision would be best to make. The role of genetics professionals is simply to provide clear, accurate information, to help people make up their own minds. This principle is particularly important when tests (not just genetic) are offered during pregnancy, the results of which can lead to some difficult choices.

How are genetic tests carried out?

A complete copy of our genes is present in almost every cell of our bodies, contained within the 46 bundles of DNA called **chromosomes**. To carry out a genetic test, the patient must first provide a sample of their DNA. This is done either by taking a small amount of blood or, for some tests, a mouthwash or swab containing a few cheek cells. In the laboratory, the DNA is examined for the presence of a particular gene change, or mutation, associated with the disease. Genetic testing can only usually be used for conditions in which the gene has been identified, so it cannot help families affected by disorders in which the faulty gene remains completely unknown. And even for conditions where the gene *is* known, genetic testing may not be possible for affected families that have a less common mutated version of the gene. Currently, genetic testing *cannot* be used to predict a person's risk of common illnesses such as heart disease, schizophrenia and diabetes, which are the result of complex interactions between many different genetic and other factors.

To carry out a genetic test during pregnancy, scientists must obtain a small amount of material from the foetus, using either **amniocentesis** or **chorionic villus sampling** (CVS). Amniocentesis, carried out at or after the fifteenth week of pregnancy, involves removing a small amount of the fluid that surrounds the foetus for biochemical or genetic testing. CVS can be carried out earlier, at or after the tenth week of pregnancy, and involves removing a few cells of the placenta. Both techniques involve injecting a needle into the womb, which can cause a miscarriage in up to one percent of cases. Because of this, researchers have been trying for some time to develop a non-invasive technique for carrying out genetic tests during pregnancy. Research into tests that use tiny amounts of "free foetal DNA" found in the mother's blood is showing promise, although at the moment it can only be used to detect foetal gender, so its use is limited to "X-linked" conditions that usually only affect boys.

Once they have a sample of DNA, scientists can identify mutations in individual genes using the techniques outlined in chapter 3. It is also possible to diagnose conditions caused by chromosome alterations, by looking at whole cells in a sample of blood or tissue. Chromosome tests are used to diagnose conditions such as **Down syndrome**, and may also reveal chromosome problems responsible for recurrent miscarriages in some women. They involve different techniques to genetic tests, and so are carried out in different laboratories. As well as detecting extra, missing or rearranged chromosomes that cause chromosomal conditions, these **"cytogenetic"** tests can also detect the chromosome alterations that are specific to different types of cancer.

Until fairly recently, the power of chromosome testing was limited to the detection of large-scale alterations, visible using light microscopes. It could not detect sub-microscopic changes such as tiny missing or duplicated parts of chromosomes. This changed with the advent of **FISH** (fluorescent in situ hybridization) technology around fifteen years ago. FISH involves the use of small pieces of DNA, tagged with a fluorescent dye. This fluorescing DNA will "light up" the corresponding bit of chromosome when examined under a special microscope, showing up any tiny alterations. A new fluorescent technique called **array CGH** (comparative genome hybridization) allows scientists to scan the entire length of all 46 chromosomes for alterations simultaneously. Although this technology is currently very expensive, many scientists think that it will eventually replace other types of chromosome analysis.

Screening programmes

Genetic tests are usually offered to families affected by genetic conditions on an individual basis. Genetic screening programmes, on the other hand, offer tests to an entire population. For example, the **"heel-prick"** test carried out on all newborn babies in the UK tests for several different genetic conditions, including **phenylketonuria** (PKU). This is an inherited condition that prevents the normal breakdown of protein during digestion, but can be very successfully treated with a special low-protein diet. Depending on where in the UK you live, the newborn blood test also checks for the presence of the blood disorders sickle-cell disease and thalassaemia, and the lung condition cystic fibrosis. Picking up these conditions as soon as possible after birth means they can be treated much more effectively, in turn delaying or preventing serious health complications. Some populations at particularly high risk of certain genetic conditions are also offered "carrier screening". For example, one in 25 people of Central and Eastern European Jewish descent is a carrier of Tay-Sachs disease, a serious condition that is usually fatal by the age of five.

All pregnant women in the UK, and many other countries, are offered screening for conditions that affect development, such as Down syndrome and spina bifida (the latter, although influenced by genetic factors, can also be the result of insufficient folic acid in the mother's diet). For Down syndrome, a combination of blood tests and a measuring of the "foetal nuchal translucency" (the amount of fluid at the back of the foetus' neck) by ultrasound can estimate the chances of having a baby with the condition. Pregnancies that have a high chance of being affected can then be tested further using amniocentesis. Since the chances of Down syndrome rise with the mother's age – about one in one hundred by age forty – many hospitals routinely offer women over 35 or 37 an amniocentesis test. However, as amniocentesis carries a risk of miscarriage of about 1 in 100 to 200 after the procedure, it is not usually offered to younger women, or those shown to have a low risk. Because **amniocentesis** can cause the loss of a pregnancy, some couples choose not to have these tests anyway, particularly if they would not consider terminating an affected foetus.

Using genetic test results

The implications of a genetic test result will depend on the genetic condition concerned, and the reasons for the test. Tests that confirm the diagnosis of a genetic or chromosomal condition allow the patients to be treated and cared for more effectively. And even for conditions that cannot be treated, a definite diagnosis can often help parents searching for an explanation of their child's health, behaviour or developmental problems. Many also cite being able to join support groups and meet families facing the same issues as another advantage of getting a diagnosis of a genetic condition. As one clinical geneticist put it, "our therapy is information".

If the test is "predictive", such as those for hereditary breast or bowel cancer, then people who find they have inherited the gene mutation can be offered frequent screening to detect the disease in its early stages. Some women at very high risk of hereditary breast cancer may opt for preventative surgery, and have a double mastectomy to cut their chances of being affected. However, there are some disorders, such as Huntington's disease, in which nothing can be done to prevent or even delay the onset of symptoms in those who inherit the gene mutation. Since there is no effective treatment for this devastating condition, only around ten percent of people with a family history of Huntington's decide to have a genetic test. Carrying out predictive genetic tests also raises ethical issues about the "right not to know", particularly if the patient is under eighteen. It is generally agreed that such tests should only be carried out on children if the results would have an immediate effect on their health care – for example, in families affected by genetic conditions that can cause death through sudden heart failure. It is not recommended for late-onset conditions like hereditary breast cancer, which does not affect women under the age of twenty – it is felt that the child should decide for themselves whether or not to get tested, when they reach adulthood.

Genetic tests that aim to identify unaffected carriers of genetic conditions can help people make reproductive decisions. For example, a woman who carries the gene mutation responsible for an X-linked condition such as Duchenne muscular dystrophy has a one-in-two chance of passing the disorder on to each of her sons. And if two unaffected carriers of a recessive condition such as Tay-Sachs disease have children, then each has a 25 percent chance of being affected. The options for such families who want to ensure they have healthy children are limited, but include using donor eggs or sperm, adoption, or terminating affected pregnancies. In the UK, termination of pregnancy is permitted up to 24 weeks' gestation, and also beyond that point, if the foetus is affected by a serious medical condition, or the mother's health is at risk. In the US, abortion is still one of the most contentious political issues, but only a very small percentage of abortions are performed after genetic testing.

Deciding whether or not to terminate an affected pregnancy can be a very difficult decision, and raises issues concerning the likely severity of a condition, and how well society accommodates the needs of children and adults with disabilities. In the end, deciding whether or not to undergo prenatal genetic testing, and dealing with the results are very personal matters that depend on each couple's individual circumstances and beliefs. The UK charity Antenatal Results and Choices (ARC; see www.arc-uk.org) provides non-directive support and information to parents throughout the antenatal (another word for prenatal) testing process.

For couples who would not consider a termination, another approach is to use **pre-implantation genetic diagnosis** (PGD). This technique involves undergoing *in vitro* fertilization (IVF – see p.148), performing a genetic test on a single cell removed from each embryo, then returning

Fertilization in the lab

Before designer babies, the newspaper headlines were all about **test-tube babies**, children conceived using **in vitro fertilization** (IVF) techniques (usually in a laboratory dish, rather than a test tube). Since the birth of Louise Brown, the world's first IVF baby, at 11.47pm on July 25 1978, an estimated three million children worldwide have been born following IVF treatment.

The procedure involves removing eggs from a woman's body, fertilizing them with sperm in the laboratory, then returning the embryos to the womb a few days later to continue developing. A related technique, called ICSI (intracytoplasmic sperm injection) involves injecting a single sperm into the egg. Louise Brown was conceived after twelve years of research by UK doctors **Robert Edwards** and the late **Patrick Steptoe**.

Her birth, at Oldham and District General Hospital, made headlines all around the world. Critics thought that this development would lead to scientists having total control over human reproduction - as in the *Brave New World* predicted by Aldous Huxley.

Since then, developments in reproductive medicine have continued to ignite controversy and attract media attention: post-menopausal motherhood, posthumous fatherhood, egg freezing, surrogate mothers, the use of donor eggs and sperm – to name but a few.

Edwards and Steptoe immediately realized that their technique could potentially be used to genetically test embryos before implanting them in the womb. However, it wasn't until 1989 that the first babies selected to be free from a genetic condition were born. UK scientists **Alan Handyside** and **Robert Winston** used PGD to select embryos free from either adrenoleukodystrophy or X-linked mental retardation – both conditions that usually only affect males.

In this case, the scientists did not test for the gene mutation directly, but looked for the presence of a Y-chromosome (females have two X-chromosomes, while males have one X- and one Y-chromosome). PGD is far more technically challenging than IVF, which itself has a relatively low success rate.

As such, it's estimated that only a couple of thousand babies around the world have been born following PGD. The evidence so far shows that children conceived using either IVF or PGD are not at increased risk of any psychological or physical health problems during childhood.

At the time of writing, the oldest of them all, Louise Brown, had just given birth to her own first (naturally conceived) baby, a boy named Cameron. However, several scientists have stressed the need for long-term follow-up studies to monitor the effects of laboratory conception on a person's long-term health.

only unaffected embryos to the womb. PGD is not widely used at present, since it is technically demanding, expensive (about £3000–5000 per attempt in the UK) and has a lower success rate than conventional IVF. But the ability to select embryos on the basis of an inherited characteristic has probably triggered more debate and media interest that any other area of genetic medicine.

Testing embryos – designer babies?

PGD is almost exclusively used by couples who have a high risk of passing on a serious genetic disorder to their children. It cannot be used for all genetic conditions, but like prenatal diagnosis, it is suitable for diseases where the single gene involved has been identified (for example, cystic fibrosis), and for disorders that usually only affect males (for example, haemophilia). PGD can also be used to avoid chromosomal conditions such as Down syndrome, by checking the number and appearance of the chromosomes in an embryo. Another use of the technique, dubbed **pre-implantation genetic screening** (PGS) or **aneuploidy screening**, has also been licensed for use in the UK. It allows doctors to select embryos with no apparent chromosome alterations, and is suitable for IVF patients who would otherwise have a very small chance of becoming pregnant. Some IVF doctors think that with improvements to technology, checking embryos in this way could one day be a routine part of all IVF treatments, but others are more sceptical.

How is PGD done?

The most commonly used method for carrying out PGD is known as **blastomere biopsy**, which involves growing IVF embryos in the laboratory for three days, at which point they consist of eight cells. At this stage, all eight cells are identical, and all are "totipotent" – that is, they still have the ability to develop into any body part or tissue. Because of this, it is possible to remove one or two cells from the embryo, using a fine glass tube, without affecting its normal development. Since all the cells are genetically identical, scientists assume that any genetic mutation found in the biopsied cells reflects the genetic make-up of the embryo. However, in

An IVF human embryo during pre-implantation genetic testing. A pipette (at left) holds an eight-celled embryo (centre) produced by *in vitro* fertilization (IVF) in the laboratory: it is being manipulated in order to have one of its cells removed. At right a smaller pipette draws off one cell from the embryo after the embryo membrane has been punctured with acid.

practice this is not always the case, which can occasionally lead to misdiagnosis of embryos.

Before being tested for a mutation, the DNA from the biopsied cells must first be copied many times, so there is enough of it to detect. This is done using the **polymerase chain reaction** (PCR), a DNA-amplification technique which, although it has been in routine use for the past fifteen years, is still technically demanding when carried out on a single cell. If the embryo is being tested for a chromosomal abnormality, then a different technique, called fluorescent in situ hybridization (FISH) is used, which allows scientists to look directly at the chromosome complement of a cell. This technique is also used to determine the sex of an embryo, by looking for the presence of a Y-chromosome.

Who uses PGD?

PGD is currently offered at only eight centres in the UK – all are fertility clinics licensed by the **Human Fertilization and Embryology Authority** (HFEA). PGD is prohibited in several countries worldwide, including Austria, Germany, Switzerland and Italy. In most countries where it is permitted, its use is limited to detecting serious medical conditions. This is the approach taken by the HFEA, which is responsible for licensing all activities involving human embryos in the UK. The US has a more permissive approach, in that PGD is unregulated. In theory, this allows clinics to provide PGD for any technically possible reason for which it is requested. In practice, clinicians in the US adhere to professional guidelines issued by the **American Society for Reproductive Medicine** (ASRM). As a result of these different approaches, the use of PGD to select babies of a particular sex is permitted in the US, whereas it is currently banned in the UK and in many other countries (see p.152).

The vast majority of couples who opt for PGD do so to avoid passing on a genetic condition to their children, or to overcome a chromosome alteration that is causing recurrent miscarriages. However, a small number of families have used PGD to ensure that a child is able to provide tissue-matched umbilical-cord blood for a seriously ill brother or sister – so-called "**saviour siblings**". The first use of PGD for this purpose was in the US, for the Nash family in 2000. Their daughter Molly, aged six at the time of treatment, has **Fanconi's anaemia**. This rare hereditary blood disease primarily affects the bone marrow, resulting in decreased production of all types of blood cells, in turn causing frequent infections and anaemia. Bone-marrow transplantation can cure many of the problems associated with Fanconi's anaemia, and a "tissue-matched" sibling is the best donor source. A transplant of cells from umbilical cord blood can also be used, providing the baby is a tissue match for the affected child.

Doctors treating the Nash family used PGD to identify an embryo that was both free of Fanconi's anaemia, and also able to provide tissue-matched cord blood cells. After four attempts, the couple conceived a healthy son, who was able to provide his sister with a blood stem cell transplant from his umbilical cord. Doctors described the procedure as "a complete success", and Molly's health was reported to have improved dramatically. The Nash family's story attracted worldwide media interest, and several families in the UK and elsewhere have subsequently had similar treatment.

Ethical objections to PGD

There are several ethical objections to PGD, some of which are rooted in a general concern about the morality of tampering with the **human embryo**, while others reflect worries over the potential misuse of this technology. The first of these objections is that PGD involves selecting and discarding some embryos, a concern that is generally articulated as "playing God", or "interfering with nature". However, embryos are also selected (and some discarded) in routine **IVF**, to identify those most likely to successfully implant in the womb. Indeed, this specific concern about PGD is generally raised by individuals and pressure groups opposed to *any* application of IVF, for this very reason.

Boy or girl?

Most people agree that terminating a pregnancy on the basis that it is the "wrong" sex is an abhorrent and unsuitable use of prenatal testing – though it is widely practised in countries with a strong preference for male children, such as China. In 2002, it was estimated that China's "one child per family" policy has led to 117 boys being born for every 100 girls. But what about methods that allow parents to choose the sex of their child *before* conception? Should they be permitted to do so? The sex of a child is determined by the sex chromosome carried by the sperm. Men produce sperm bearing either an X- or Y- chromosome, whilst women produce eggs that all contain one X- chromosome. If a sperm bearing an X- chromosome is united with the X from the egg at fertilization, this will result in an "XX" pregnancy that produces a female baby. If a sperm bearing a Y chromosome fertilises the egg, an "XY" pregnancy will result, giving rise to a male child. Using *in vitro* fertilization (IVF) with pre-implantation genetic diagnosis (PGD), early embryos can be tested to determine which sex chromosomes they carry and, consequently, either male or female embryos can be selected for implantation. This technique is mostly used by couples who want to avoid passing on diseases caused by a faulty gene on the X chromosome, which usually only affect boys. But there are also a small number of UK couples who have sought to use PGD to determine gender for non-medical reasons – so-called "social sex selection".

Social sex selection is currently not permitted in the UK, since the **Human Fertilization and Embryology Authority** (HFEA) will only issue licences to clinics to carry out PGD for medical reasons. This ban was challenged, unsuccessfully, by the Masterton family in 2000. Alan and Louise Masterton wanted to use PGD to have a baby girl, after their only daughter died in a bonfire accident. The couple, who have four sons, eventually travelled to a clinic in Italy to have the treatment, where the procedure was at that time permitted, but gave up after three unsuccessful attempts.

Another objection to the use of PGD is that it is an inherently eugenic procedure that raises the spectre of Nazism. But there is evidently a big difference between a state-controlled effort to eliminate certain types of living people, and personal reproductive decisions aimed at having a healthy child. In any case, both PGD and prenatal diagnosis can only be used for families identified as having a family history of a genetic condition, or individuals in certain high-risk groups. Even if all these couples chose not to have affected children (which they don't), many genetic conditions arise as a result of new gene mutations in the egg or sperm – about a third of all cases of Duchenne muscular dystrophy, for example – so affected children are often born to parents who have no family history of the disorder. This fact alone would make genetic testing either before or during pregnancy a fairly ineffective eugenic procedure. However, there *is* a real concern – expressed most often by those involved in the disability

In 2003, **Nicola Chenery**, mother of four boys, gave birth to a baby girl after receiving PGD treatment at a Spanish clinic. Despite these highly publicized cases, the demand for such treatment in the UK has remained low, and studies show that most people using PGD for sex selection is permitted in the US, where it is available commercially. So is another procedure called sperm sorting, provided by a firm called Microsort. This technique involves separating fresh sperm on the basis of whether they carry an X or a Y-chromosome. The success rate of sperm sorting – about 91 percent for females and 76 percent for males, according to the company's website – does not match that of PGD (almost 100 percent), although it is much cheaper, and less invasive.

Some ethicists condone the use of sex selection for "family balancing" – couples who already have one or more child of a particular gender seeking to have a child of the opposite sex – but not for choosing the sex of a first child. Others argue that while many people may disagree with the use of reproductive technology for social sex selection, to ban it outright in a democratic society is an unacceptable curb on an individual's liberty. Arguments against using PGD for this reason include the concern that it will lead to a societal gender imbalance, that it is inherently sexist (neither of which applies to couples seeking gender variety) and that it represents a "slippery slope" into a world in which babies are increasingly viewed as consumer objects. Such children, it is argued, will not be valued for themselves, and will carry an unacceptable weight of parental expectation to behave in a way that befits their gender. However, some parents will always have certain expectations of their children, rightly or wrongly, and it is not clear how banning sex selection addresses this issue. But it seems likely that even in countries where social sex selection (and other non-medical traits, should the technology ever make it feasible) is allowed, most people will continue to leave reproduction to chance, rather than face the inconvenience, uncertainty and cost of assisted reproduction techniques.

rights movement – that offering PGD (and prenatal tests) could lead to a devaluing of individuals with disabilities, and shift resources away from their care to programmes aimed at preventing their births. Clearly, a compassionate society should take steps to ensure that this does not happen.

Finally, some people are concerned that the use of PGD to select embryos free from disease is a step down the slippery slope to "**designer children**", chosen for attributes such as height, intelligence, and sporting ability. However, such traits, like many common illnesses, are the result of many different genetic and non-genetic factors. It will probably never be possible to select embryos with a "shopping list" of desired characteristics, even if there were ever a demand for such children. The one exception to this is gender, a non-medical trait for which some couples have already used PGD to select either a boy or a girl.

A brief history of eugenics

Eugenics gets a bad press, and quite rightly too. Words such as "Nazis", "racist pseudoscience" and "never again" spring to mind. Yet the early eugenicists claimed that their intentions were good. What was it in their ideas that went so wrong? The word eugenics, from the Greek for "good in birth", was coined in England by the Victorian polymath **Francis Galton**, Charles Darwin's cousin, in 1883. Galton had been pondering *The Origin Of Species*, with its new emphasis on inheritance, and argued that much human misery was caused by physical problems which were passed down the generations. He wondered whether these misfortunes could be avoided, and speculated as to whether breeding from the better specimens would produce healthier, happier people – maybe even smarter ones too (like him).

The idea that we might be able to boost the quality of the stock of the next generation by paying attention to our choice of mates had a wider appeal. It was an extension of the Enlightenment project to improve life before death. If you supported education and public health, then it seemed to many to be a logical, more direct route to enhancing the human lot. There were a couple of problems, however. First, eugenics did not just mean encouraging parents of the right "quality" to have more children (**positive eugenics**). It also took in discouraging the less promising specimens of humanity from breeding (**negative eugenics**). Galton began by emphasizing the possibilities of promoting good marriage, but also wrote at length about the need to curb breeding among the inferior, which he often defined by their race or class. The whole enterprise depended on knowing what qualities were actually "good", and which really were inherited, as opposed to the results of people's upbringing.

This was where a welter of fears and prejudices entered in. The teeming underclass of the newly industrialized Victorian cities were, in the eyes of eugenicists, poor specimens – ones who threatened to outbreed the gentry. In the US, the fear of "degeneration" fixed upon immigrants from

Eastern and Southern Europe and non-whites. And when the ideas of Mendel's genetics (see p.40) came into fashion in the early twentieth century, a long list of ill-defined traits – such as feeble-mindedness – featured in eugenic propaganda, aided by "research" from institutions such as the **Eugenics Record Office** at Cold Spring Harbor in New York.

For some, eugenic goals justified state-enforced sterilization, or even euthanasia. Both had their advocates in the US. According to **Ewin Black**, author of the well-documented and chilling *War Against The Weak* (2003), "various methods of eugenic euthanasia – including gassing the unwanted in lethal chambers – were a part of everyday American parlance and public debate some two decades before Nevada approved the first such chamber for criminal executions in 1921".

Francis Galton

While euthanasia was mostly just talked about, sterilization was actually a widespread and long-lasting practice in the USA. Forced sterilization laws were passed in over thirty states, beginning in 1907 in Indiana. Some, notably California, used them regularly in the 1920s and 1930s. Tens of thousands of people, mostly inmates of state-funded institutions for the mentally handicapped or mentally ill, were sterilized.

American and German eugenicists had something of a mutual admiration society in the first decades of the twentieth century. The Model Eugenical Sterilization Law, published by the US Eugenics Record Office in 1914 provided for sterilization of the "feeble-minded, insane, criminalistic, epileptic, inebriate, diseased, blind, deaf; deformed; and dependent" – including "orphans, ne'er-do-wells, tramps, the homeless and paupers". The policies enacted in Germany when **Adolf Hitler** came to power in 1933 took inspiration from them, and then some. Eugenic philosophy did not lead directly to genocide, but it helped it along – and so did many German geneticists.

The full story of twentieth-century eugenics is much more complex than this, however. There were socialist, Zionist eugenicists in Weimar

Wendell Holmes and negative eugenics

The argument for negative eugenics to relieve the supposed burden of bad genes on the community was ringingly endorsed by Justice Oliver Wendell Holmes, handing down the US Supreme Court's decision in 1927 to uphold Virginia's eugenic sterilization law. The test case was **Carrie Buck**, a teenage mother who was supposedly feeble-minded like her mother before her. She was to be operated on entirely justly in Holmes' view. It was a principle of public health.

"We have seen more than once", he wrote, "that the public welfare may call upon the best citizens for their lives. It would be strange if it could not call upon those who already sap the strength of the State for these lesser sacrifices, often not felt to be such by those concerned, in order to prevent our being swamped with incompetence. It is better for all the world if instead of waiting to execute degenerate offspring for crime, or to let them starve for their imbecility, society can prevent those who are manifestly unfit from continuing their kind. The principle that sustains compulsory vaccination is broad enough to cover cutting the Fallopian tubes ... Three generations of imbeciles are enough."

He based his judgment on witness testimony about the character and mental capacities of Carrie, her mother Emma – an inmate of the Virginia Colony for the Epileptic and Insane – and her baby daughter Vivian. The Colony Superintendent Dr Priddy advised the court that "these people belong to the shiftless, ignorant, and worthless class of anti-social whites of the South".

Research in the 1980s revealed that Carrie was neither promiscuous – her illegitimate daughter was the consequence of her rape by a relative of her foster parents – nor feeble-minded. Her daughter Vivian, who died when she was only eight, performed no worse than her fellow students in the little time she spent at school. But both were condemned by the court on the basis of flimsy evidence and flawed assessments. Four thousand more involuntary sterilizations were carried out in the same Lynchburg hospital where Carrie Buck had her operation, the latest as recently as 1972. They included her sister Doris, who who was operated on without her knowledge while in hospital for appendicitis, and who regretted until the end of her life that she never had children.

You can read the late Stephen Jay Gould's impassioned essay on the new facts of the case at www.stephenjaygould.org/library/gould_eugenics.html

Berlin, and progressive, Fabian eugenicists in Britain. The British were less keen on negative eugenics, although a bill introduced by **Winston Churchill** advocating compulsory sterilization of "the feeble-minded and insane classes" – as he called them – was only narrowly defeated in 1913 by a combination of Leftist and Catholic opposition. There were extensive, and only recently widely known, sterilization efforts in Sweden, that bastion of social welfare, and eugenics movements in South America and Asia.

But it is the atrocities committed in Germany in the name of racial purity which mean that the few who today label their plans for the future of humanity "eugenic" know they are going out on a limb. Even those devoted to positive eugenics, the only kind likely to be even remotely acceptable as a topic of discussion following World War II, tend to be treated as somewhere between eccentric and sinister. The Repository for Germinal Choice, the self-styled "Nobel sperm bank" owned by the late US multi-millionaire **Robert Graham** is a case in point. Operating in California from the late 1970s until 1999, it managed to produce just over 200 babies using sperm from donors who fit Graham's rather narrow criteria for super-citizens. A few, including the brilliant physicist **William Shockley** – a man scarily obsessed with "degeneration" – were actually Nobel laureates. Others were more run-of-the-mill achievers in business or athletics. None were notable artists or humanitarians. For the stories of several of Graham's donors, and some of their offspring from artificial insemination, check out David Plotz's enterprisingly researched *The Genius Factory* (2005).

More broadly, the fact that there are still a few racially inspired eugenicists

Carrie Buck – a victim of enforced sterilization in the 1920s (see opposite)

upholding the faith on the Internet shows that really bad ideas can survive even the moral lesson of genocide. But where does that leave modern

Genetics and eugenics, old and new

When eugenics became a dirty word after World War II, geneticists were at pains to argue that their work had nothing to do with the pseudo-scientific approach to human traits that they claimed characterized the excesses of the eugenicists. And most of them denied that there was any sense in applying genetics to try and maintain or improve the quality of the population.

This separation of eugenics and genetics was generally accepted, despite the close involvement of some prominent geneticists with eugenics before the war – and not only in Germany – and the continued insistence of some key figures that the eugenic project was valid. The Nobel Laureate **Herman Muller**, for example, remained an advocate of positive eugenics throughout his life, as well as raising concerns about the effects of mutations that might be caused by nuclear fallout in the 1950s.

In practice, too, the development of medical genetics in the second half of the twentieth century was, on close inspection, always linked in with eugenic ideas. The first modern-style **genetic counselling clinics** ("heredity clinics" as they were known in the US at the time) were run by clinical geneticists committed to "non-directive" advice. But when patients asked, "Doctor, what would you do if you were me?", the answers they got tended to be clear. In the early days, sterilzsation might be offered after couples had a child affected by a genetic disease.

A pioneering genetic clinic was set up in Michigan in 1940, by **Lee Dice**, later President of the American Society for Human Genetics. In his 1951 address to the fledgling society, then three years old, he made it clear that the Nazis' programme was repugnant. But he was still sympathetic to sterilization of "defectives" – the old eugenicists' favourite target – under "proper safeguards".

But any programme of sterilization big enough to eliminate any large proportion of harmful genes would infringe liberty, he conceded. His alternative was to persuade people not to reproduce.

attempts at human betterment that revolve around improving genes? Are they eugenic? And does that matter?

The American historian **Diane Paul** has argued that the E-word is no real use any more. It is quite often heard in discussion of new genetic technologies – with regard to **prenatal screening** for diseases like cystic fibrosis, for example. But what most often happens is that people with different views still agree that eugenics is unacceptable. So if they oppose some new application of genetic technology, they claim it has eugenic tendencies. If they support it, they say it does not. Like the journalist's "designer babies", eugenics is a shorthand for things we do not want to accept – a word to end an argument.

This was the assumption underlying the "non-directive" advice in Dice's clinic. A similar approach was taken by **John Fraser Roberts**, who set up the first genetic counselling clinic in Britain at the Great Ormond Street children's hospital in 1946. Although Roberts was a long-standing member of the British Eugenics Society, he advocated non-directive counselling based on risk estimates.

Dice wanted to see a national movement: "The knowledge of genetics which already has been acquired through research is largely unavailable to our people... I urge, therefore, the establishment in every state of a series of heredity clinics which will cooperate closely with physicians, dentists, hospitals, schools, probate judges, welfare agencies, and others responsible for the public welfare, in order to provide dependable advice on human heredity."

This is the thinking that fed into modern-day genetic counselling, which strives not to be eugenic. Whether it is always non-directive in practice still depends both on what counsellors say to their patients, and on how they are heard, but that is certainly the governing professional ethic. These days, genetic counselling is an integral part of medical genetic testing services. Genetic tests can be carried out for a variety of reasons: to confirm a diagnosis, to identify healthy "carriers" and to predict who might be at risk of a genetic condition (see chapter 7 for more on genetic testing). Modern genetic counsellors talk to patients and their families about the implications of a genetic test result, and can provide information on risk to other family members. In 2003 the UK Department of Health published a White Paper promising £50 million to develop genetic services in the National Health Service, including £18 million for genetic testing in specialist centres and training fifty genetic counsellors. The workload of genetic testing laboratories has increased greatly in recent years, with the identification of more disease genes, particularly those that cause hereditary cancers. There are now around 25 such centres throughout Britain (for details see the British Society of Human Genetics website, www.bshg.org.uk). In the USA, the National Society of Genetic Counsellors has well over 2000 members (see www.nscg.org).

But is it? Philosopher **Philip Kitcher**, British-born but long resident in the US, thinks it is more complicated than that. In his book *The Lives To Come* (1995) he argued that, "Once we know how to identify the genotypes of future people, eugenics is the only option." He did not mean that the **Human Genome Project** would bring on a new round of "fitter babies" contests, or eugenic sterilizations. From him it was an objective matter of logic, following by default from his definition of eugenics. According to him, "A eugenic practice is an attempt, by some group of people, to shape the genetic composition of their descendants according to some ideal." So if you know how to affect the gene pool in more or less controlled ways, then whatever policy you adopt – even not using that knowledge at all

– is taking a view on what kind of people there will be. Strictly speaking, that is eugenic.

Kitcher says there are four factors to consider in the assessment of any eugenic measures:

▶ The **subpopulation** whose reproductive activity is to produce the desired results

▶ The degree to which **they make their own (reproductive) decisions**

▶ The **characteristics** according to which choices are made

▶ The **quality of the genetic information** used in making the decisions

The eugenic policies that any sane person rightly deplores plainly occupied a very bad place on the map defined by these four factors. But there might still be other places on the map that would be worth exploring – such as the possibility of individuals being able to freely choose to use highly accurate genetic information to make their own decisions about gene variants that everyone agrees it would be good to avoid.

That does not stop the arguments, of course. Nor should it – these are very important issues with far-reaching consequences. Kitcher admits that his fully informed and freely chosen reproductive selection is an ideal unlikely to be realized by his term for it: "utopian eugenics". How free can decisions be when some people have expert knowledge about genes, while others are struggling to understand the terminology? Or when prospective parents know all too well the limits on social support for sick or disabled children? How good are the predictions of any particular genetic test? And how will the effects of many individual decisions influence the choices of those who come after? Free markets can have consequences nobody likes in much less important areas than human genetics.

These points crop up regularly, but the arguments about what currents run in the gene pool have some new issues to deal with. These have arisen as the technology was invented to do more than the old-style eugenicists ever imagined – increasing the contribution of some good genes, diluting some less good ones. What if we can actually construct better genes (in some sense) and introduce them into some humans' genetic make-up? This tends to be discussed not under the heading of eugenics, but "**enhancement**". And we will do the same, but not until chapter 15.

My DNA made me do it
Genes and the human condition

How strongly can DNA influence complex characteristics? And does the writ of DNA govern human beings as strongly as other living things? These questions have generated huge controversy. They follow on from the Victorian debate over man's place in Nature, and whether we are risen apes or fallen angels. They evoke echoes of worries about the fate of the **germ plasm** (see p.86). And they arouse political passions stoked by sex, race, and class, and philosophical concerns about free will, and the nature of good and evil.

A **molecule** is not the cause of these problems, nor the answer to them, even if it is the molecule which carries the information in our genes. But it is implicated in lots of claims about why people do the things they do. Most of the science covered in this guide so far has focused on its impact on health and medicine, for the simple reason that most human genetics research aims to tackle diseases. But some believe that genetics will eventually also help answer bigger questions about what it is to be human. This chapter will look at how genes might influence the way we behave, and at attempts to explain how certain traits may have been selected during human evolution – whether they affect social behaviour (**sociobiology**) or individual mental abilities (**evolutionary psychology**). There's a crucial distinction to bear in mind here. Genes and behaviour studies are about difference – what makes an individual more likely to behave in a certain way while another less so. Sociobiology and its more recent offshoot, evolutionary psychology, are about similarity: evolved aspects of human nature which we may all share.

Genes and human behaviour

Let's start with the differences between people. Newspaper headlines in recent years proclaiming the existence of the "gay gene", or hinting that a weakness for alcohol could be in the genes more or less implied that everything we do is somehow genetically predetermined. But this is much too simple a view of human nature. So exactly what role do our genes play in shaping the person we are? No modern scientist seriously believes that there is a gene for criminality carried only by people who break the law, or that there is a gene for addiction.

But genes may influence your personality, and it may be that people with certain types of personality are slightly more prone than others to such behaviour. There is also evidence that different versions of certain genes – for example, those that make proteins involved in the breakdown of **toxic chemicals** by the body – can affect the liability of becoming addicted to alcohol or drugs. Yet environment and upbringing play an equally important role, and in many cases a decisive one. A behavioural pitfall such as alcoholism is the result of neither nature nor nurture, but a complex interaction between the two. We discuss some of the general reasons for this later in this chapter.

Most of the research in this area has focused on the factors that influence addiction, in the hope of better understanding – and so better treating – alcoholism and drug addiction. But scientists have also probed the extent to which personality and **sexuality** are inherited, with some intriguing, if inconclusive results. A lot of this work has involved studies of twins, especially **identical twins** (who share identical genes) separated at birth, because they can potentially shed light upon which aspects of a person's behaviour are learnt, and which develop from their biological make-up. Unfortunately, studies of adoptees can't easily tease apart the effects of genes from other factors that affect a foetus whilst it is growing in the womb, although they can provide an estimate of how biologically "innate" a characteristic may be. And doubts have also been cast on the validity of studies that compare identical with **non-identical twins** (see chapter 4 for more on twin studies). Despite these limitations, a handful of studies linking specific genetic variations to particular traits seems to back up the idea that our genes can influence (rather than determine) certain aspects of our behaviour.

Dangerous knowledge?

Searching for genes involved in behaviour is controversial, especially for those genes that may influence aggressive or addictive behaviour. Understandably, some worry that this knowledge may be misused to discriminate against people, or provide excuses for antisocial behaviour. There are also fears that in the future, parents will make misguided attempts to genetically "improve" their children by getting rid of "undesirable" characteristics before they are born. Such concerns are not unfounded (although the last of these is probably beyond the realms of scientific possibility), since studies of human genetics have been tragically misused in the past. Followers of the **eugenics** movement (see chapter 7), at its height in the early part of the twentieth century, believed that some people were genetically superior to others. They wanted to prevent people they regarded as being genetically inferior from having children. As a result, many in the US and other countries were forcibly sterilized, a practice that only stopped in the 1960s. Eugenic policies also helped pave the way for the Nazi atrocities committed during World War II. Eugenics was based on prejudice and presumption – its followers assumed that all human characteristics were genetic, and ignored the possible effects of other factors, such as standards of living, diet or upbringing.

A more recent example from the "old" genetics of the temptation to oversimplify links between genes and behaviour was the supposed association between males with an **extra Y-chromosome** and criminality. In 1965 a survey of inmates in a prison in Scotland found that seven out of the 197 prisoners with a history of violent crime had an extra Y-chromosome. As the normal incidence of XYY males is roughly one in a thousand, this led to press reports of a "criminal chromosome". On the other hand, XYY males tend to be taller than average, and may be more susceptible to mild learning difficulties, so perhaps any XYY criminals are more likely to end up in jail than their XY brethren, because they are just more likely to get caught. And proposals to test the link, via prospective studies of XYY newborns, were an ethical minefield of course. Who was going to be told what about these babies? When were they to be told? And how might this affect their upbringing?

Later studies found no link between XYY and violent crime. Nonetheless, a number of abortions were carried out in the US on fetuses which were incidentally diagnosed as XYY after **amniocentesis** (a common prenatal test involving a small sample of amniotic fluid). More generally, there are still those who believe that some people are born superior to others, and scientific evidence is still open to misinterpretation and misuse, as much as ever. However, the dangers of genetic discrimination need to be weighed against the potential health benefits of modern genetics research – most of which seeks not to "improve" the human race, but to advance the health and choice of the individual. For example, understanding human behaviour may help develop new ways of treating addiction, anxiety and eating disorders.

Genes and personality

Your experiences, upbringing and circumstances are probably the most important influences on your personality. However, your genes may affect how anxious you are, or how easily you become depressed, for example. In turn, this may affect how you react to certain situations. But genes are far from the whole story – although identical twins share identical genetic information, they often have very different personalities. And even if scientists identify which genes are involved, say, in aggression, it probably won't help predict who is going to behave in an aggressive way. With these caveats, what follows is a brief summary of what we know so far about how genes affect personality.

Aggression

Scientists have identified several different brain chemicals that can affect your mood. Some people may inherit variants of genes that alter the levels of these chemicals, all types of neurotransmitter, which could make them more easily annoyed. In 1993, Dutch scientists found a family in which several men seemed prone to unusually aggressive behaviour, with many convictions for arson, rape and attempted murder between them. At least eight had a variant of the **MAOA gene**, which makes a protein called **monoamine oxidase A** – which in turn helps control levels of brain chemicals such as serotonin (the brain's "mood" chemical). The affected men had abnormally low brain levels of these chemicals, which the researchers suggested might have caused their aggression.

Although such families are few and far between (thankfully), a later joint UK/New Zealand study found that children who had inherited a particular version of the MAOA gene were more likely to be antisocial, defined this time as showing increased tendencies towards bullying, lying or flouting school rules. However, they also discovered that a genetic predisposition to antisocial behaviour is heavily influenced by upbringing – children who inherited the MAOA variant grew up to be model citizens if they were parented well. And those who had been abused seemed to cope much better with the abuse if they had the more active form of the MAOA gene. Similar results have been found in monkeys deprived of maternal care. This serves to illustrate that although nature – MAOA (and other genes that affect serotonin levels) – may play a role in aggressive behaviour, nurture definitely has the last word. As several scientists have pointed out, since ninety percent of murders are committed by men, the biggest genetic risk factor for aggression is having a Y-chromosome. But as nobody has yet been excused their violent behaviour as an unavoidable side effect of being male, it's unlikely that many judges will see fit to allow genetic excuses for bad behaviour in the future (though some accused have tried this defence – see p.166).

Taking risks

Risk taking is another personality trait that, according to studies of twins, appears to be influenced by our genes – although nobody has any real idea which ones. "Novelty-seeking behaviour"– a personality classification used by psychologists which can be tested by questionnaire – covers everything from a penchant for extreme sports to gambling or experimenting with drugs, all of which can trigger an adrenaline rush and increased levels of the chemical **dopamine** – the brain's main pleasure chemical. All animals have this natural reward system, which is usually triggered by food or sex. A widely reported 1996 study linked the possession of a "long" version of the **dopamine D4 receptor gene** (D4DR) to novelty-seeking behaviour – the theory being that this genetic variation made people less sensitive to the effects of dopamine, making them more prone to seek bigger thrills to achieve feelings of satisfaction. However, as often happens with gene and behaviour studies, attempts to replicate this finding have produced mixed results, suggesting that several other genes and non-genetic factors are involved in thrill seeking. Research into links between DRD4 and attention deficit hyperactivity disorder is ongoing, and likely to have more practical applications. As well as risk taking and aggression, various studies have found evidence for a genetic influence on traits such as shyness, happiness and sociability. There is also extensive research on anxiety, in both humans and rats and mice, but again early reports of strong single-gene effects have given way to analyses which indicate there are a large number of genes involved, each contributing a small amount to the "nature" side of the equation.

Genes and addiction

Some people, it seems, are more likely to become addicted to alcohol, nicotine or other drugs than others, because of the influence of their genes. It seems that roughly speaking, the more addictive a drug is, the greater the role played by genetic factors, with cocaine and opiates such as heroin being amongst the most addictive of substances and also the most "heritable" addictions. This suggests that there is an underlying biological basis

A genetic defence?

Can you blame your genes for your actions? In 1992, Californian lawyer **John Baker** did just that. Affected by alcoholism, he began to drink heavily when his business ended up in debt. He embezzled money from a client's account, and when caught, he initially pleaded guilty to the crime. But his legal advisor suggested he change his plea, because John Baker's father – of North American Indian descent – was also an alcoholic. There is a high incidence of alcoholism in Native American communities, which some researchers have suggested may be due to an increased genetic risk. To their joint surprise, the defence worked.

But a similar approach by lawyers acting for **Stephen Mobley**, on trial for the murder of pizza-shop manager John Collins in 1991, failed to impress the judges. Mobley shot and killed 24-year-old Collins during a robbery in Georgia, US. His lawyers cited the 1993 Dutch study on the MAOA gene to argue that Mobley, whose family had a history of aggressive behaviour, had a genetic predisposition to violence. The defence was not successful – the court argued that "the theory of the genetic connection is not at a level of scientific acceptance that would justify its admission" – and Mobley was eventually executed in March 2005.

for some types of addiction, inherited variations in which can increase or decrease individual risk. Drugs hijack the brain's reward system, triggering the release of dopamine in the absence of food or sex. However, the effects soon wear off, resulting in cravings for a return to the pleasurable state induced by the drug – a cycle that leads to addiction in some people. Some of the genes that appear to affect a person's risk of addiction are involved in controlling levels of dopamine in particular parts of the brain, while others affect serotonin.

Alcoholism

The propensity for alcoholism to run across generations in families has been hotly debated. Might it be because children inherit genes that increase their risk, or because they learn their drinking behaviour from their parents? Several different genes are undoubtedly involved. For example, some genes affect how well you hold your drink. People who break alcohol down quickly can drink more, and so are more at risk of alcoholism. The biological basis for these differences lies in variations affecting genes that make **liver enzymes**, which convert alcohol to **acetaldehyde** (alcohol dehydrogenase, ADH), and then into **acetate** (aldehyde dehydrogenase, ALDH). A build-up of **toxic acetaldehyde** – caused either by a fast-acting ADH or a slow-acting ALDH enzyme – triggers an unpleasant "flushing reaction" that discourages further drinking. In several eastern Asian countries, such as Japan, most of the population has either one or both of these gene variants, which effectively affords them genetic protection against alcoholism.

Other clues to genetic influences on alcoholism have come from work on animals. Alcohol affects animals in the same way it affects people: it slows them down. A team of US researchers studied the effect of alcohol on different strains of mice, and found that some strains had more severe withdrawal symptoms than others. It seems that at least four different mouse genes are responsible for these differences, and that human versions of these genes may be involved in alcoholism. Other researchers are studying the effects of alcohol on fruit flies. But animal studies may be of limited use – a gene that does one thing in an animal may do something completely different in humans. Furthermore, our behaviour is far more complex than that of animals – mice and flies do not have stressful jobs, or peer pressure to cope with, for example.

Some of the genes involved in alcoholism may influence personality, which in turn could influence attitudes to alcohol. A 1993 study of personality traits found that alcoholics and their relatives tended to score higher in "harm avoidance" and "novelty seeking" – summed up by one researcher as the "type of person who wants to sky-dive but is afraid of the plane". But genes are far from the whole story – availability of alcohol, depression, stress, upbringing, circumstances and peer pressure may all influence the risk of alcoholism. It is unlikely that genetic research will ever lead to preventative measures to stop people becoming alcoholics, but many researchers hope that shedding light on the biological processes involved in addiction will result in better, more targeted treatments. In the meantime, anyone with a strong family history of addiction to alcohol should be aware that if they start drinking heavily, they may find the habit very hard to kick.

Genes and sexuality

Gay men are more likely to have gay brothers, uncles and cousins than heterosexual men are – especially on their mother's side of the family. Data on twins also indicated a genetic influence on sexuality, although genes cannot be the whole story – even with identical twins, the chances that one twin will be gay if his brother is are around 50/50.

In 1993, a group of US researchers led by **Dean Hamer** of the US National Cancer Institute reported that gay men tended to have a particular marker on the X-chromosome (the chromosome men inherit from their mothers) in common with their gay relatives. However, a larger study in Canada reported six years later (which received less media coverage than the original claim) failed to replicate this finding, and no "gay gene" has ever been found. Despite this, there is some evidence to support an underlying biological basis for homosexuality, though it's unclear whether this is the effect of genes, other factors that affect the developing foetus (for example, female hormones), or an interaction between the two. For example, the neuroscientist and gay activist **Simon LeVay** has reported finding differences in one crucial, small area of the brain, part of the hypothalamus, when he examined post-mortem samples from gay and straight men. But as well as extended controversy about the validity of his findings, which were first published in 1991, there remains the question of whether the difference in behaviour could affect the brain or vice versa.

Little research has been done on female homosexuality, partly because it is believed to be less common than the male equivalent. This is obviously a difficult subject to research, but most surveys indicate a lower incidence of same-sex experiences in women. Again, twin and sister studies point to a genetic influence, although, as for males, any genes involved presumably affect the chances of being bisexual, rather than homosexual. Otherwise, they would never have been passed on from one generation to the next – unless they conferred some other evolutionary advantage. Like work into genes and personality, however, studies of sexuality have not attracted much of the funding poured into genetic research, so many of these questions remain unanswered.

Are genes us?

The role of genes in making us behave differently looks set to go on generating controversy for some time to come. But then their role in determining human nature as a whole – the things we all have in common – is equally debated.

Debates about human nature stretch back over millennia of philosophy and centuries of science. But we are just going to mention two sets of claims heard in the last few decades. First came **sociobiology**, a sweeping attempt to account for social behaviour using evolutionary theory. Then came **evolutionary psychology**. The two are closely related, although there are crucial differences. Both make intriguing claims that will be either appealing or alarming according to your attitude. Both have also been criticized for making arguments that go beyond the evidence. The problem in both cases relates to many of the claims discussed in the first half of this chapter about genes and behaviour. It seems to lie in the gaps of a story that has to start with genes, but which ends up a long way from what genes definitely do – preserving information which is read out and interpreted at the molecular level.

Let's start with sociobiology. The term, widely publicized after the great US biologist **Edward Wilson**'s *Sociobiology* in 1975, followed from Wilson's claim that the social sciences are best viewed as part of biology. It was a reaction against a widespread view in sociology and anthropology following World War II that human nature is, if not a "blank slate", then enormously flexible. We have a biological nature, of course, but it is our nature to be strongly shaped by learning and culture. In this view, humans live in a realm of language and symbol that evolves independently and much more rapidly than genes and DNA.

Sociobiology brought together two main sets of ideas that pointed in the other direction. One came from studies of animal behaviour, in the hands of pioneering ethologists such as **Konrad Lorenz** and **Niko Tinbergen**. Lorenz, in particular, clothed the idea of instinct in modern dress by showing how many of the ways in which birds and animals behave – in courtship and mating, for example – follow fixed patterns that are species-specific. Mostly, the evidence suggests, they are inherited, not learned. When the picture is more mixed, as when Lorenz's newly hatched ducklings "imprinted" on him – they tried to follow him around as if he were a mother duck – it is interpreted as an inherited tendency which can produce varied results in practice. Normally the first large, moving object that a duckling sees is its own parent. So being, in effect, programmed to follow it has survival value.

Lorenz's "instincts" were ill defined, although he also had a lifelong concern with genetic decay, which he believed was caused by domestication of animals and civilization of humans. His early attachment to Nazism did not stop him being awarded a Nobel prize with Tinbergen in 1973 – though this was more for his techniques of animal observation than his actual theories.

The other, more important stream feeding sociobiology was directly related to genetics. Evolutionists had long puzzled over examples in the natural world of what, in people, we would call **altruism**. An animal

Konrad Lorenz takes a new year's day stroll in 1973

sounding an alarm call when it spies a predator, for example, warns the rest of the herd, but the animal making the noise may well end up getting eaten. You would have thought that an animal would just slink away and hope the hungry visitor will satisfy its appetite on someone else.

The puzzle was that if a gene can influence this behaviour, selection ought to lead to silence. A mutation which makes a preyed-upon animal even a little more likely to lie low and stay silent ought to improve that animal's chances of passing genes to its offspring, over one which sounds the alarm.

But selfish genes can lead to unselfish behaviour if they affect your relatives. The British geneticist **J.B.S. Haldane** famously remarked that he would lay down his life for two brothers, or eight cousins. The idea this sums up nowadays goes by the name **inclusive fitness**, a term put forward by another Englishman, Bill Hamilton, in 1964. It leads to a calculus that shows how selection can still favour a gene that promotes behaviour which compromises the individual's chances of survival. Not surprisingly, the theory helped biologists to understand many features of the collective lives of social insects such as ants, bees and wasps – Wilson's specialty. Thousands or millions of workers, soldiers or drones are sterile, and yet they maintain a queen who sends their genes on to the next generation.

Put together ethology and **selfish-gene theory** – that term would be coined by Richard Dawkins the year after Wilson's book came out – and you have the beginnings of the rather wider-ranging notion of sociobiology. Wilson synthesized a vast amount of knowledge from study of different kinds of creatures, from corals to ants to monkeys. But his real message, in the first and last chapters of his book, was about humans. We, too, have species-specific patterns of behaviour which are universal. And they exist because they evolved in ways that adapted humans better for survival.

Like most such grand visions, there was something in this. The first claim sociobiology makes is surely true. Genes enable us to do what we do, but they also constrain what we can do. Human beings can walk upright on two legs, but cannot fly. And if our bodies are shaped by evolution, then so are our minds. If Darwin was right, then our qualities of mind emerged because they conferred reproductive advantage. Inclusive fitness shows, in principle, how the ways we act toward each other can be linked directly to the influence of genes.

Accept all that, though, and there is still lots of scope for disagreeing about how many aspects of our nature, expressed through our minds, are fixed by evolution, and how strong the influence is. Humans must do the same things as all other creatures – survive and reproduce, and so are linked with all the rest of the living world in the great game of natural selection. But we are also exceptional. We have language and culture (and write books). Just possibly, our genes, building our minds, evolved ways of escaping from some of the constraints that other living things have to put up with. The fact that they will never be conscious they are putting up with them defines the difference.

How important that difference is remains a matter of opinion. No one really holds either of the extreme positions – that the human mind is a blank slate on which culture can write anything at all, or that most of our behaviour can be tracked back directly to inbuilt tendencies governed by our genes. But there are people with strong leanings either way, and they tend to be impressed by different kinds of evidence. Wilson, in the end, is impressed by ants.

But however much observation of how different creatures act lends itself to explanations of why these actions aid their survival, can we fill in the rest of the story? That story must connect genes to (human or ant) behaviour. Why? Lots of things can affect an individual's chances of surviving, and the number of offspring it leaves behind. But from a biological point of view the genes are the crucial thing which pass to the next generation. So selection must act, in some way, at their level.

That means that when it comes to behaviour, there must be some link between the behaviour and the genes if selection is going to keep them going. But, if you have read this far, you will recognize that the part of the

A Darwinian dust-up

Sociobiology nowadays sits quietly on the shelves of university libraries everywhere. But it was one of the most controversial books of the decade when it appeared in the middle of the 1970s. It was the central volume of a trilogy that opened with *The Insect Societies* in 1971 and finished with *On Human Nature* (and a Pulitzer prize) in 1978. The first book conveyed Wilson's immense knowledge of ants, bees and the like. The third expanded *Sociobiology*'s remarks about humans to book length. The basic message was that there is a fixed human nature, common to all, arising from our evolutionary development. Ant nature arises because ants need to behave in certain ways to carry out their survival strategies. And so does human nature.

Sociobiology was seen at the time either as a heroic synthesis of practically everything there was to know about evolution, population biology, genetics and behaviour (the central 550 pages or so) or a provocation to right-thinking people everywhere (the opening and closing chapters).

The controversy was partly academic turf war. Wilson started with the book proffering a hope that "not too many scholars in ethology and psychology will be offended by this vision of the future of behavioural ecology". But by its close he was offering the grand vision he returned to in *Consilience* at the end of the 1990s, in which "the humanities and social sciences shrink to specialized branches of biology; history, biography, and fiction are the protocols of human ethology; and anthropology and sociology together constitute the sociobiology of a single primate species".

This annoyed lots of academics in other disciplines. But Wilson's strongest critics were fellow evolutionists such as American leftists **Richard Lewontin** and **Stephen Jay Gould**. Their attack was intensely political. They reckoned that Wilson's argument about evolved constraints upon how humans behave, and upon what we can become, was mere ideology. The complex and often heated arguments that followed are related in a number of later books, notably Andrew Brown's popular *The Darwin Wars* and historian Ullica Segerstrale's monumental *Defenders Of The Truth*.

Wilson himself was humane and liberal, as well as a brilliant man, so looking back it is odd to recall him facing an onslaught from people who shared his basic conviction about the importance of evolutionary theory. But his extension of biology to features of human behaviour, and human societies, as well as chapter titles like "The Morality of the Gene", did leave himself open to a kind of guilt by association. No claim about human nature is tied tightly to any political position. But we do live in a society in which people like to invoke biology to explain why some kinds of behaviour will never go away. As Kenan Malik observes in his subtle review of the issues in *Man, Beast And Zombie*, "nature" is a bit like God – it always seems to be on your side. Whenever you read an article "explaining" why men don't do housework, or conjecturing that the "glass ceiling" preventing women rising to the top of big corporations could ultimately stem from differential rewards for risk taking in hunter-gatherer cultures, the reasoning is rooted in sociobiology.

story that explains how a gene (or a change in a gene) can alter behaviour is largely still a blank. There are fully worked out stories for some simpler organisms – the **nematode worm**, say, or the **fruit fly**. And there are findings in humans which seem to offer some kind of proof of principle. Injecting **testosterone** seems, among other things, to make people more inclined to aggression. Presumably a gene that promoted higher levels of testosterone might, other things being equal, do the same thing.

Even here, however, with a "simple" hormonal effect, there are many contradictions resulting from studies of **brain development**, **neurotransmitters**, and the effects of **social position**. And in most cases, the intricacies of the connections between genes, brains, minds and behaviour are beyond us. This leaves us not so much with a blank slate, but a blank canvas, on which people can paint the picture they want, as long as the viewer does not ask about the details. What is more, when it comes to the rather large class of features of human societies that are cited in *Sociobiology* as universal – from religious ritual to incest taboos – there is always the alternative explanation that they were invented as part of culture and passed on by learning.

Stone age Darwinism

Evolutionary psychology is sociobiology's younger brother. As its name suggests, it focuses more closely on the mind. It looks more at what might have been the case in the past as evidence for human nature, as opposed to sociobiology's inventory of present-day customs. And it is more interested in psychology than behaviour. The favourite metaphor in the field is that the mind is a bit like a Swiss army knife, with different tools – or modules – for different kinds of task. The ideas about how it works are a mixture of evolution and computer science. For the leading evolutionary psychologists **Leda Cosmides** and **John Tooby** of the University of California, Santa Barbara, "the mind is a set of information-processing machines that were designed by natural selection to solve adaptive problems faced by our hunter-gatherer ancestors".

Again, all this is pretty plausible at some general level. Yes, the mind did evolve. And we know that different parts of the brain get involved in different activities. There are even brain cells, **neurons**, which specialize in a single task – such as the neurons in the visual cortex that recognize the sensory input from a horizontal line, for example. But how far does this go? We just do not know. Cosmides and Tooby's units of selection seem to be not genes but undefined bits of neural tissue they call "**brain circuits". As they say, "Generation after generation, for 10 million years, natural selection slowly sculpted the human brain, favouring circuitry that was good at solving the day-to-day problems of our hunter-gatherer ancestors". One objection to this idea is that it assumes that the places humans lived stayed the same all that time, and that we have not had a

Can genes control brains?

So can genes control brains? Well, they can preserve a programme that guides the development of the cells that make up brains, and some of their responses to their surroundings. But the brains they help build – in creatures which have an organ worth calling a brain – are complex and highly flexible. And human brains, immersed in culture as well as ecology, are the most flexible of all.

This general argument is not new. One way of summarizing it is to say that, even if genes are hard-wired, brains need not be – a sort of computer-related metaphor which seems to fit the general presuppositions of many writers about genes and behaviour. That is a bit out of date because it is now clear, as much of this book describes, that genes are not hard-wired, either. A tight coupling between a particular **single DNA sequence** and a fixed outcome in the cell or the organism is the exception, not the rule. It is even less the case in our heads. What we know about development of the brain suggests that the generation and location of **neurons** is influenced by genes, but other processes then ensue which are much more independent. This applies, to some extent, to the connections between neurons (those "circuits"), which are continually changing throughout our lifetimes as part of the processes of learning. Earlier on, it applies to which ones survive. We now know that more than a third of all the neurons born in the brain die off by the time a child is three. There is a vast pruning of cells and connections which, the evidence suggests, is mostly influenced by the kinds of experiences the infant is having. There may well be evolutionary processes going on, but they are in play among populations of neurons in a single brain, not among genes.

Another way to look at the issue is to try some numbers. Counting all the cells in a grown human is harder than counting the number of different **gene sequences**, but estimates range up from ten million million to around ten times that number. We now reckon there are about 25,000 different genes. Some are active in all cells to maintain their basic functions. The rest switch themselves on and off at different times, depending on the type of cell, and so far biologists have listed just over two hundred different kinds of human cell. With that kind of diversity, the problem of accounting for how cells know what kind of cell they are, and where in the body they are supposed to do their stuff, seems intuitively manageable with some clever combinations of genetic and chemical signals, without worrying about how complex the details are going to be. Of course, a vast amount of cell biology is devoted mainly to working out those details, and lots of interesting problems remain, but there's no reason to think the job cannot be done.

chance to evolve since our environment changed. Neither of these things is necessarily true. But if they are, then perhaps this sculpting fashioned parts of the brain which are specialized for things such as reading others' emotions, deciding which potential mates are attractive, using tools, or speaking a language grammatically. In some of these cases we have a

Brains, however, are more complex still. The average **neuron count** in a human brain is around 100 billion (thousand million), the textbooks say. On examination, each neuron has something between 1000 and 10,000 connections to other neurons. Somewhere in all those connections are the "neural circuits". And the governing logic of the evolutionary view of the mind says these must be altered over time in ways which are subject to natural selection. But we have remarkably little idea about what the kinds of circuits assumed to operate here are like, or how they might work.

Quite a lot is known about what the different parts of the brain look like, and about the way quite small regions are specialized, and are active during particular tasks. The evidence comes from **brain injuries** (old-style neuroscience) and from the spectacularly sensitive images that come from **magnetic resonance** and other new kinds of scanner (new-style neuroscience). But knowing, in effect, that circuits in some part of the brain light up when it is doing a particular thing – whether seeing, speaking, or scheming – tells you nothing about how the circuits actually operate.

Finally, it is worth considering the real neural circuits that have actually been studied. They are typically in much simpler organisms, mostly **invertebrates** (lobsters for instance), and involve a bunch of **nerve cells** that can be teased out from the whole creature and analysed in detail in the lab. How many cells? A million? A few tens of thousands? Well, no: more like a couple of dozen.

So the scale up from here to the neural circuitry which must be in some way involved in the things that lobsters or fruit flies do not do – learning a language, say, designing a dwelling, or negotiating a complex social structure – is more than a bit of a stretch. That is no reason not to study models for the interactions between genes and neural circuits, and to look at how they might evolve together. But it is a reason to be cautious about extending what we know about these much simpler systems to human brains in all their wondrous complexity. You may feel, like us, that the problem of understanding all the many influences on behaviour – genetic, environmental, cultural, accidental – how they relate to precise details of connections in the brain, and the ways those connections give rise to the intricacies of each person's behaviour, remains rather mind boggling. If you do not, you are probably ready to sign up to do some research in evolutionary psychology.

rough idea which large-scale brain regions may be involved. But, as with genes and behaviour, the parts of the story which would tie genes to brain circuits are basically not written yet. Whether they ever can be written is still a matter of opinion.

Meanwhile, the programme of evolutionary psychology begins to look overblown compared with what it actually delivers. Look in detail at the insights that evolutionary psychology offers into behaviour, and the claims tend to appear softer. They are about tendencies rather than certainties. Take one of the EP notions for which the evidence is strong – that step-parents are more statistically likely to abuse their stepchildren than biological parents to abuse their own genetic offspring. Canadians Martin Daly and Margo Wilson call this the **Cinderella Effect**. And they have statistics to back it up. In countries where these things are recorded systematically, children are hundreds of times more likely to suffer abuse at the hands of step-parents than genetic parents.

As well as confirming the bad rep stepmothers get in fairy stories, this fits with a genetic story well enough (although we have no idea which genes are involved). Perhaps human males – the main perpetrators of abuse – want to eliminate their mate's children by a previous partner, it is postulated, to ensure any resources going benefit children who bear their own genes, as happens in some other animal species. Or perhaps spending days with a squalling baby makes almost any adult have murderous moments, and the restraints on acting on such a dreadful impulse are not as tightly bound if the child is not genetically your own.

Either way, it is not clear how much use knowing this information is. Doubtless it would be of importance to social workers if step-parents may statistically present a higher risk to the average child. But even if this was true, they would still need to judge individual cases in the knowledge that the vast majority of step-parents are not child murderers. So evolutionary psychology's big insights seem to have rather small payoffs.

PART 4
GENETIC
TECHNOLOGIES

Modifying genes

GM technology, GM politics

The discovery of the restriction enzymes which bacteria use to chew up foreign DNA, along with other enzymes which can paste bits of genes back together again, was the key ingredient in the first practical method for genuine genetic engineering. As we described in chapter 3, since the mid-1970s it has been possible to alter a species – not by breeding, or by exposing it to radiation or some other mutagen and hoping for some useful result – but by taking a gene out of one creature and inserting it another.

This recombinant DNA approach was pretty hit-and-miss at first. Sometimes it still is. But, as in other areas of science, biotechnology has moved on a long way since then, and if you really want to move a gene from one place to another, you can now usually get it there. It may even work, and perhaps even do what you expected it to. In this chapter, we review some of the often controversial results of applying this new-found ability to real projects.

Making medicines in bacteria

Human insulin

The first important real-world application of recombinant DNA trickery was to make an old product in a new way. The story of **recombinant insulin** shows some of the problems of getting new genes to work, and of the early operation of the **biotech industry**.

All people with Type 1 diabetes (the kind that usually starts when you are young), and many people with advanced Type 2 diabetes (the kind that usually begins in middle age) need **insulin** shots to manage their blood sugar. The injections replenish levels of a tiny protein hormone which should be made by small clusters of cells in the pancreas. People with Type 1 diabetes no longer make insulin, while those with Type 2 have become resistant to its effects. The stuff used to be extracted from pigs or cows, which works – although it can provoke allergic reactions in people sensitive to the small differences between the kinds of insulin. However, in the 1970s, the main provider, **Eli Lilly and Company**, was looking for new sources. Type 2 diabetes was (and still is) on the rise and demand was rising faster than they could easily meet. Maybe genetic engineering was the answer.

There were three groups in the race to be first to programme the E. coli bacterium to manufacture the human protein. The two main contenders were led by **Walter Gilbert** at Harvard, signed up to biotech company Biogen, and **Herb Boyer** and **Stan Cohen** from Stanford, working with the new company **Genentech**. They chose quite different approaches to the problem – which involved getting hold of the DNA coding for human insulin, inserting it into E. coli, and making it work.

The difficulties arose from some of the complications of normal, naturally occurring insulin production. In human **pancreatic cells**, a rather large **messenger RNA** has some non-coding intron sequences removed. It is then transcribed into a long chain of amino acids, which go through two more stages of processing to produce the two short, cross-linked chains of active insulin. Bacteria, with their normally much more simply organized genomes cannot do any of this.

To give them something they could work with, the Gilbert group decided to use DNA copied from mammalian mRNA. They worked first with rats, fishing the correct RNA out of preparations of pancreatic cells, then using the enzyme **reverse transcriptase** to make single-stranded DNA, and **DNA polymerase** to double up the DNA strands. Hey presto: a new gene for insulin.

They then had to stitch the gene, and a suitable promoter sequence, into a suitable **plasmid** (a DNA molecule capable of self-replication) together with markers indicating when the whole package had been taken up by a transformed bacterium. In principle, this only needs to happen once, and the power of bacterial replication does the rest. By supplying nutrients, you soon have a dish, or even a vat, of E. coli expressing the gene for insulin. In practice, in those early days, it was all immensely demanding of technique

Genentech

Any company with "gen" somewhere in the name probably dates from the early days of the biotech boom in the late 1970s and early 1980s, such as Biogen, Amgen, Regeneron, and Synergen. (Sadly the Southern European biotech start-up GenItalia exists solely as a favourite joke among biotechies.) It was an investment spree not quite on the same scale as the later dot.com bubble, but one similarly fuelled by wild enthusiasm for new technological possibilities.

It started small, when an ambitious 27-year-old entrepreneur called **Bob Swanson** met Herb Boyer in 1976. The result was the first of the new-style biotech companies, **Genentech**. When the company went public on Wall Street in September 1980, just after an agreement with Eli Lilly to manufacture human insulin, the $35 opening price rose to $89 on the very first day. It was the fastest share price rise ever seen. Boyer and Swanson both became paper millionaires.

Most fledgling biotech enterprises failed, or were swallowed up by other companies. But Genentech had staying power, and later products of theirs included bacterially produced human growth hormone, a natural blood clotting chemical for haemophiliacs, and the breast cancer drug (actually a monoclonal antibody) **herceptin**. Swanson died in 1999 but the company went on growing. It currently has a market value of $100 billion, and its revenues in 2005 were more than $6 billion. Valuable stuff, this DNA, when you know what to do with it.

You can tell that Genentech were in the game right from the start by the address of their (very informative) website: www.gene.com

– as science journalist **Stephen S. Hall** relates in his excellent book *Invisible Frontiers*. The job was even slower because the freshly coined rules written by the US National Institutes of Health for DNA manipulation meant that when Gilbert's team moved on to human DNA they had to fly to England to do the work at Porton Down – where they could get access to a germ warfare lab with the high-level containment required in the early days of recombinant DNA for mixing genes between species.

Their rivals Boyer and Cohen were spared the highest containment the NIH specified because they thought it would be easier to do the job using synthetic DNA. Their plan was to custom-make DNA sequences that coded for the two insulin chains one base at a time – the whole protein had been sequenced by **Fred Sanger** back in the 1950s, a feat which won him his first Nobel prize. They would then make two different recombinant plasmids in similar fashion to Gilbert, and transform two bacterial lines so each would make one of the two insulin chains. The plan was to recover the two protein chains from the cultures and mix them in the right conditions for them to fold up and link together as normal.

Walter Gilbert, co-founder of Biogen and joint Nobel-winner (with Fred Sanger) in 1980

There were plenty of problems along this route, too, but Boyer and Cohen, and Genentech, won this particular race. Recombinant insulin produced using their method was brought to market by **Eli Lilly** in 1982, and remains a valuable product. There have, however, been persistent reports that the new product gives a few diabetics less warning of low blood sugar (**hypoglycemia**) than bovine or porcine insulin. Some patients have resolved this by switching back to the animal-derived hormone, which remains available.

So far as producing medically valuable products in bacteria goes, Gilbert's **cDNA method** works better with larger genes and is now the industry standard. It has been used to make a large array of proteins and smaller peptides, including hormones, clotting factors, and immune system agents such as **interferon**.

The recombinant DNA debate

In July 1974 a group of top biologists published a letter urging their colleagues to down tools. The members of a Committee of the US National Academy of Sciences feared that new experiments that involved swapping DNA might be dangerous. They wanted to call a halt so that the hazards could be properly assessed.

Their proposal for an experimental moratorium still seems extraordinary. Molecular biologists were eager to exploit the new advances that had recently been made to manipulate DNA. Chief among them were a recently discovered class of bacterial enzymes, known as **restriction enzymes**. These enzymes, evolved for defence against foreign DNA from viruses, chop up DNA into small pieces, but always at the same short base sequences. This meant that they could be used to "cut and paste" new bits of DNA into the small circles of free-floating DNA found in many bacteria known as plasmids. In effect, genetic engineering was now really possible for the first time. DNA molecules from different sources – or different species – could be cut, spliced and rejoined, or recombined, and grown in the common lab workhorse *E. coli*. A fantastic new world of experiments beckoned in which the workings of genes from more complex organisms would be revealed by transplanting them into this simple bacterium.

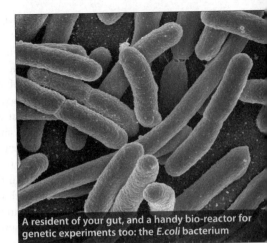

A resident of your gut, and a handy bio-reactor for genetic experiments too: the *E.coli* bacterium

But in those pre-genomic days great swathes of the DNA sequences of higher organisms were poorly understood. Suppose nasty things lurked in the unexplored regions? And there were more specific worries about the spread of **antibiotic resistance** – normally carried on plasmids – or even genes that could cause cancer. The letter (known as the Berg letter, after the committee chair, future Nobel laureate **Paul Berg**) called for a halt to two kinds of experiment: those which would introduce new antibiotic resistance, or the genes for making new toxins in bacteria; and those giving them new genes from animal viruses.

A scientific conference at Asilomar in California in February 1975 debated the hazards, with a carefully selected band of journalists allowed in, provided they did not file reports until the end of the discussion. After lengthy disagreements, the assembled scientists took up an idea from **Sydney Brenner**. Back home in the UK, Brenner had been forcing down milk laced with the normal lab strain of *E. coli*, known as **K12**. He found that the bug did not colonize the human intestine – well, not his anyway. Why not go one further, he suggested, and develop a strain of the bacterium which could not survive outside the lab? That would be sufficient assurance for most experiments. Any which might pose a more serious hazard could be conducted with extra, physical precautions. On this basis, the Asilomar group recommended that the moratorium should be lifted.

That eventually happened, with their ideas incorporated in official guidelines from the **US National Institutes of Health** in 1974, and broadly similar official regulations elsewhere. But not without a struggle. The politics were fraught. To scientists, the techniques were incredibly enticing – they are now the staples of hundreds of thousands of labs. They were anxious to go ahead, and to be allowed to regulate themselves. But the possibility of new risks drew lots of attention. If they were going to take all these precautions, some reasoned, this stuff must really be dangerous. And the revolutionary potential of the techniques excited both biotech investors and groups worried about the wider implications of genetic engineering. Some leading scientists tried to confine the discussion to purely technical issues. Would the right mix of safety measures guarantee to keep the new organisms in the lab? We already knew how to contain dangerous pathogens – health research and germ warfare had seen to that. Applying such knowledge to the containment of weedy little bacteria, which ought not to survive outside their carefully formulated nutrient broth, should have been enough to satisfy anyone, reasoned those in favour of unrestricted research. So long as everyone signed up to the proper precautions, there would surely be no problem.

Yet this was not enough to reassure the critics – eco-activists such as the American lawyer **Jeremy Rifkin**, radical scientists like biology Nobel laureate **George Wald**, and microbial ecologists who worried that the gene-splicers rarely encountered the complexities of how organisms behave outside the lab. Do we really *trust* researchers in a highly competitive field to go by the book, without laws to force them to? And what about all the other issues raised by fiddling with DNA? As well as the long-term consequences of altering various species – perhaps even the human species – there were plenty of other arguments to consider. In the years following the rise of the environmental movement, and protests against the Vietnam war, some wanted to discuss bigger questions such as whether there was a right to free inquiry, responsibility in research, public participation in decisionmaking about science, who was likely to benefit from new work, and whether stronger corporate ties to university labs were desirable.

In the USA, and to a lesser extent in Britain and elsewhere, the round of official committees, Congressional hearings and local inquiries into recombinant DNA which ensued was paralleled by an unofficial debate. This discussion, largely aired in the media, saw fears of Frankensteinian experiments, accusations of "playing God", and references to the *Andromeda Strain* movie all used to put the issues into a more dramatic frame.

As now, however, these were the kinds of fears that were hard for legislators and regulators to address. They were more at home with the

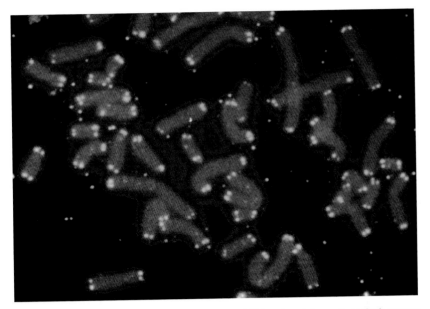

Above: telomeres are the vital "buffers" of repetitive DNA and protein coating which protect the ends of chromosomes. They gradually shorten over our lifespan.

Below: the enzyme telomerase (which elongates telemeres) is extracted from cancer tumour cells for experimentation. Many cancer cells are considered "immortal" because telomerase activity allows them to divide virtually forever.

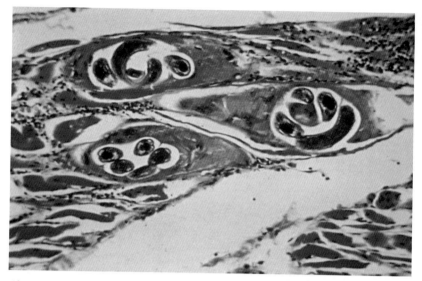

Above: a tiny roundworm (*Trichinella spiralis*), within human muscle tissue. It is a parasite transmitted by eating undercooked pork, and its genome was one of the first entire genomes to be sequenced.

Below: these three little pigs are transgenic: they were genetically altered to produce Human Protein C. The latter is used in blood production, and could have applications as a very effective blood-clotting agent.

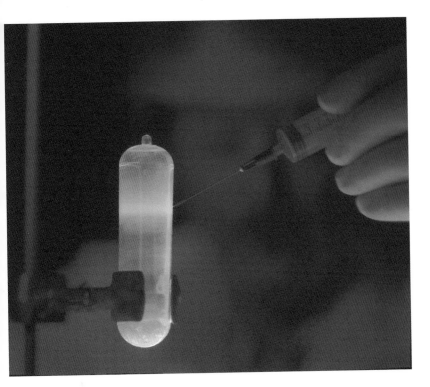

Above: purified DNA is extracted to be used for molecular biology studies. The DNA binds a dye which absorbs UV light and makes the DNA fluorescent orange.

Right: a mouse cDNA microarray containing approximately 8700 gene sequences. It reflects the gene expression differences between two different mouse tissues. DNA microarrays (also known as a gene chips are collections of microscopic DNA spots attached to a solid surface, such as glass, plastic or silicon chip, forming an array for the purpose of expression profiling, monitoring expression levels for thousands of genes simultaneously, or for comparative genomic hybridization.

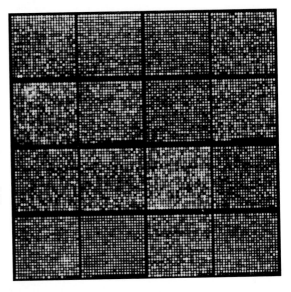

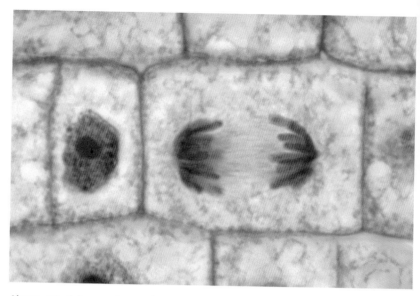

Above: mitosis (here at the anaphase stage) in a plant cell (an allium root). Mitosis is the process by which the a cell separates its duplicated genome into two identical halves. Chromosomes, replicated and condensed, line up and divide along an axis.

Below: a close-up of the first metaphase – in which homologous pairs of chromosones align – of meiosis in cells of a member of the lily family.

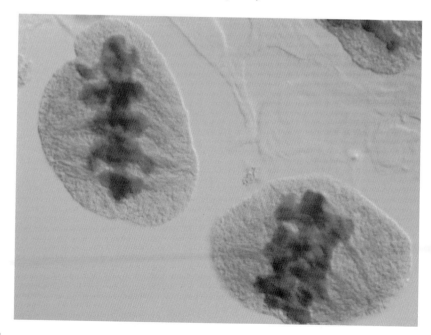

more technical issues of hazards and protection, and the debate gradually returned to the original questions about laboratory risks and safety rules. There would be no self-regulation, but the moratorium was followed by an official system for assessing the possible risks of an experiment, and matching them to a level of containment. This impeded some research – there was a spell in the mid-1970s when some important experiments could only be done in the few places equipped for germ-warfare research. But as the work progressed, it appeared the hypothetical risks were just that. The guidelines were gradually reduced, and the techniques of recombining DNA became routine. By the mid-1980s the original regulations in the US and UK, for example, had been largely dismantled, and the UK's **Genetic Manipulation Advisory Group** had been completely wound up.

So was the original ban an overreaction? Many scientists thought so, not least **James Watson**. The double-helix hero was one of the signatories of the Berg letter, but came to regret it. He was irked by what he saw as inconsistencies in the Asilomar recommendations (mammalian DNA inserted into bacteria warranted high containment; frog DNA did not). And he was enraged by some later opponents of recombinant DNA research, calling them "an odd coalition of spaced-out environmental kooks and leftists who see genetics as a tool for enslaving the masses".

But this is too simple a view of them. They, and the politicians, were responding to the feeling that the direction biology was heading in would need some kind of overview which took account of wider social interests. The whole episode was part of a process that is still ongoing, in which everyone tries to figure out how to do this. Is it possible to get the benefits of the technologies that read and rewrite genes while minimizing the risks, whether those risks are environmental, social, or ethical and moral hazards?

Thirty years on, with the technologies several more leaps forward, we are still working on it. There is a huge pile of academic studies of the recombinant DNA discussions – the weightiest of all by the US historian Susan Wright in her book *Molecular Politics*. They are a rich resource for understanding the new politics of science.

Transgenic animals

Recombinant DNA technology is applicable, with the right additional tricks and tweaks, to any organism – not just bacteria, but plants, animals, even humans. We look at plant and animal uses in agriculture later in this chapter, and at human genetic modification in chapter 14. Here, we review some of the more indirect applications in medicine – and one rather more direct one: modifying animal organs for transplantation.

Many genetically modified animals are created for medical research, to provide models for human disease. Mice, rats and, more unexpectedly, **zebra fish** (not that much like humans, but still vertebrates) have all become laboratory fixtures, but mice are way ahead, so we will concentrate on them.

Mice have been bred on an industrial scale for science for many years, particularly for cancer research, but the combination of gene modification technology and the genome projects has confirmed their central place in the lab. Around 99 percent of human genes have a mouse counterpart, mice multiply quickly, and GM procedures tend to work well in mice.

There are efforts afoot to produce a complete set of "knockout" mice – rodents which have a single gene that is inactivated – to study gene function. Meanwhile, there are many laboratory strains of mice already in use in work on particular diseases. These tend to be artificial "products" because natural models of serious human disease are rare – they are normally selected out. Early examples of informative mutations included an alteration in the enzyme **glucokinase**, which leads to diabetes in mice and humans, and a mutation known as **shaker1**, which leads to hereditary deafness.

More recently, there has been lots of important work done using mice with gene alterations that replicate aspects of brain diseases such as **Parkinson's**, **Alzheimer's** and **Huntington's**. As with other conditions, mice are used both to help understand what causes the symptoms, and to test out possible therapies.

Although the use of animals in labs has fallen in recent decades, all this means that use of GM mice has increased. In 2004, the proportion of animal experiments involving genetically modified animals rose from 27 percent in 2003 to 32 percent in 2004, although the figures include many mice that are simply used for breeding. Views about this differ largely, in line with views about animal experiments in general. Those who are happy to eliminate mice from our kitchens and cupboards using traps, poison, or just by keeping a cat, are generally unlikely to be troubled by their other life in the lab. Some would say that, if we accept the necessity of their use to humankind in a laboratory, then increasing that use by genetically modifying them is but one simple step.

What different people are willing to accept, however, is a moot point. The announcement of the first GM primate, **Andi**, in 2001, raised fears of a similar rise in the use of primates in medical labs. Andi was a rhesus monkey whose cells had incorporated a jellyfish gene when monkey eggs were exposed to a modified virus. But the team at the Oregon Regional Primate Research Center were testing the technology in the hope of creating primate models of human disease. Also troubling to many is the notion of taking organs from animals to keep humans going. The pig is the main research target

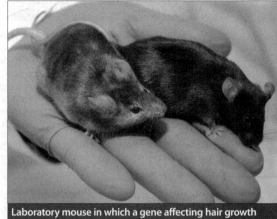

Laboratory mouse in which a gene affecting hair growth has been knocked out, left, next to a normal lab mouse

here, because it is roughly the same weight as a human and its organs are the right size, too. Unfortunately, as well as being notably intelligent, it also harbours a large population of old viral genes which have become incorporated into its cells' DNA. These so-called **endogenous retroviruses** abound in higher organisms, and are mostly harmless **information parasites**. But they *might* not be, and the minute but alarming possibility that transplanting a pig organ into a human could allow a dormant retrovirus to run amok in a new host appears to have stalled serious proposals to move to trials of xenotransplanted hearts or kidneys, at least for the time being.

Down on the pharm

A final use of GM animals is for **drug manufacture**, rather than research. In the earlier development of biotechnology, a popular scenario was that suitable mammal species would be modified so that they made medically valuable products in their milk. This, it was argued, would be easier – and cheaper – than using cell cultures, especially if the cells had to be from higher organisms rather than bacteria. Much research was directed to this aim, including the programme which produced **Dolly the sheep** (see chapter 10).

However, bringing products made this way, by "**pharming**", to market has proven much harder than for drugs or hormones made in GM bacteria. The enthusiasm triggered by the birth in 1990 of **Herman** – a trans-

Why xenotransplantation?

When curative medicine falters, we already replace some body parts when we can. Sometimes, if all you need is a relatively small piece of tissue, it can be replaced by taking it from somewhere else. Hence the popularity of the heart bypass, which improves blood flow to the heart by taking a stretch of artery out of the patient's thigh and grafting it into their chest. But that isn't usually possible. We don't have a surplus of most kinds of tissue. And since we don't yet know enough biology to grow a new one just where you need it: if you really need a new heart, for example, your only hope is a transplant.

Where from? The options are sadly limited. Your doctor can try to get hold of a human heart which the previous owner has no further use for. This more or less works, but the secondhand heart market is seriously bad at supplying them. You can find some ingenious technologist to make you a heart-sized machine out of metal and plastic – the thing is only a pump, after all. Promising, but not yet an effective substitute for the real thing. Or, just maybe, you could use a living, non-human heart.

In that light, if you are lying around with a serious heart condition, waiting for the final flutter, the pig might look promising. It has a heart about the same size as a human, pumping a similar volume of blood for an encouraging number of years. And, if you are happy to tuck into bacon and sausages, you are unlikely to have major qualms over the pig's say in the use of its heart. But biology has placed an annoying obstacle in the way of this simple scheme. If you just take a pig heart, and transplant it into a human (no matter how skilful the surgery) the patient's immune system will recognize it as foreign tissue and destroy it impressively quickly.

However, now that it is possible to perform genetic surgery as well as the more traditional kind, there may be a way round the extreme immune reaction to an organ from another species. A key to this reaction is the recognition of certain special chemicals on the outside of the pig's cells – so-called **cell-surface antigens**. Precisely which antigens get put there is under genetic control, and the genes are tolerably well understood. So if you could fix the pig's genes so that it didn't make the tell-tale antigens or, even better, made human ones instead – you could grow a pig whose organs would be effectively "disguised" and ought to transplant successfully into a person.

The possibility of viral transfers is reason for caution, but the transplant donor shortage is getting worse. Even when we have two of the requisite organ, living human donors remain reluctant. There are 2000 heart transplants annually in the USA, less than one percent of the number of deaths from heart failure. Similar shortages face prospective liver and kidney recipients.

genic bull who was supposed to sire cows who made a key human protein, **lactoferrin**, in their milk – has largely fizzled out. One big setback was the failure of the Scottish company **PPL Therapeutics** in 2003. PPL, which was linked with the Roslin Institute group who raised Dolly, hoped to sell the anti-emphysema and cystic fibrosis drug **alpha-1-antitrypsin**, made in sheep's milk. The company got as far as signing with pharmaceutical giant **Bayer** to run Phase 3 clinical trials – the last stage before a drug is approved for prescription – in 2000. However, early results identified some patients whose lung problems were worsened by the drug (or its aerosol delivery) and the trial was stopped.

This was bad news for the sheep as well as the patients. After Bayer pulled out, PPL was crippled by the costs of maintaining a flock of expensive GM sheep that had to be constantly monitored for infection. Most of the sheep – 3000 in Scotland and another 3500 in New Zealand – were slaughtered and incinerated.

However, there have been some recent successes in the field that suggest it is too early to write off the potential of pharming. GTC-Biotherapeutics in Massachusetts now has a licence to market an anti-clotting agent Atryn, made in the milk of transgenic goats. And other companies have turned to rabbits as living factories, with some success. At the moment, "**big pharma**" have little involvement in the area, but with biopharmaceuticals worth more than $30 billion annually in the US, there is plenty of money to develop animal use further if and when it proves advantageous.

GM foods

Altering the genetic make-up of plants has provoked almost as much heated debate as has tampering with the DNA of animals or people. Much of the fiercest controversy around genetic modification in recent years has concerned crop plants. Maybe it is because, unlike drugs made in genetically altered bacteria, the plants are out in the world for all to see. And food is symbolically important. We care about what we eat, and what we feed our families. So how far has genetic engineering got in plants, and what are the issues?

All the familiar crop plants are the result of old-school genetic modification, through traditional breeding, of course. This is the foundation of modern human culture, the technological leap that created the agricultural surpluses essential for civilization. But this kind of selection and judicious hybridization, even when rethought in a modern genetic framework, can only improve crops by reshuffling genes that are already present in closely related species.

Since the recombinant DNA pioneers assembled the tools of modern genetic engineering, it has been possible to fix up plants with new genes from anywhere you like. The techniques differ in detail because it is a bit harder to get new DNA into plant cells, but the principles are the same. In theory, a DNA sequence from any species can now be inserted into the genome of a plant, and the gene can be made to work, though most crop modifications still involve genes from other plants. In 2004, the USA had by far the largest acreage given over to growing GM crops, with around 50 million hectares of GM soybeans, maize, cotton and canola. Argentina, Canada, Brazil and China also had appreciable GM acreage, but the amount of GM cultivation elsewhere, especially in Europe, remains very small, largely because of public misgivings about messing about with genes in foodstuffs.

The modifications that have been sold to farmers so far by the seed companies using new technology are mainly about yield rather than quality or enticing new varieties. With a few exceptions, existing GM crops which have found widespread use are not about irresistibly fragrant and luscious new fruits, for example. They are sold as aids to the farmer in fighting insect pests and eliminating weeds.

The most famous insect-fighting aid is the use of genes from a type of bacteria first discovered during investigation of a disease that was killing Japanese silkworms in 1901. It was later named **Bacillus thuringiensis** and found to be one of a class of bacteria that produce **crystalline insecticides**. Thousands of varieties are now known which can produce over two hundred different insecticidal crystals. They found a use in insect control, but they were difficult to apply and not very long-lasting compared with the synthetic chemicals widely used by the mid-twentieth century.

However, "Bt" looked more attractive by the 1980s, with increasing resistance to synthetics – from both the insects themselves and environmentalists. Bt is organic, targets specific pests, and degrades after use. In the GM era, the idea that plants could make Bt crystals directly seemed appealing. No spraying would be needed, and insects which attacked the plant at any stage in its growth would ingest the crystals – lethal to them but harmless to humans. An organic insecticide would be co-opted for the latest hi-tech approach to agriculture. Genetically engineered corn with Bt genes on board was first marketed by Monsanto in 1995, and potato and cotton followed. All three are in wide use, with competing commercially available strains coming from other companies like Novartis (formerly CIBA) and Aventis. Each is engineered (or "enhanced" as the companies tend to say) to make a Bt toxin targeted at specific pests which attack the crop in question.

When the farmer's problem is weeds rather than insects, the strategy is indirect. Take a plant-killing chemical, and then make the crop you want to harvest resistant to it. Then you can spray your fields and watch all the other plants die off. The leading example is the herbicide **glyphosate** – or **Roundup**, as it is commercially known – the global bestseller first introduced by Monsanto in the mid-1970s. It works by inhibiting an enzyme in a pathway many plants use to make essential amino acids. Without the right amino acids to make proteins, the plant withers and dies. But if a crop plant is sprayed during a growth period, it is harmed too, which means the chemical can normally only be used to clear ground of weeds at certain times of year.

Turning to GM, **Monsanto** first set out to boost production of the enzyme, known as EPSP synthase, in chosen plants, by incorporating extra copies of the gene taken from petunia plants. However, this only gave partial resistance to glyphosate, so they tried again with an alternative form of the enzyme found in bacteria which are largely immune to the effects of the herbicide. "Roundup Ready" strains of soybeans and cotton incorporate this gene together with other DNA sequences that control where and when the enzyme is produced in the plant. They are now widely used in the US and elsewhere. They increase herbicide use because farmers now douse the fields instead of spraying round the edge. Farmers who buy their **Roundup Ready** seed have to sign an agreement not to keep any for replanting, or to pass it on, so they become regular customers for both the seed and the herbicide. Eighty-nine percent of the US soybean crop was made up of varieties engineered for herbicide resistance in 2006. Other Roundup Ready crops now in use include maize, canola (known as oilseed rape in Europe), and sugar beet, with rice, wheat, potatoes and even Roundup Ready trees on the way. These are the modifications that have been widely used, and discussed and argued over in great detail. But they are very much products of the first stages of genetic modification. So far there has been little effort to improve the actual product, but that will come, as we discuss below.

What are the problems with GM?

The advent of GM crops has generated debate about a whole raft of issues. They tend to be posed all at once, as Britain's **Prince Charles** did when he asked "Ten Questions on GM Food" in the *Daily Mail* newspaper in 1999.

▶ Do we need GM foods in this country?

▶ Is GM food safe for us to eat?

▶ Why are the rules for approving GM foods so much less stringent than those for new medicines?

▶ How much do we really know about the environmental consequences of GM crops?

▶ Is it sensible to plant test crops without strict rules in place?

▶ How will consumers be able to have real choice?

▶ If something goes wrong with a GM crop, who will be held responsible?

▶ Are GM crops really the only way to feed the world's growing population?

▶ What effect will GM crops have on the world's poorest countries?

▶ What sort of world do we want to live in?

That last, open-ended question suggests that the agenda here is pretty wide. But the directly relevant GM crop questions can be split into four kinds. Could they harm human health? Could they damage the environment? Who will they benefit? All those three relate to a final big question: how should GM crops be regulated?

Each of these raises subsidiary issues. On health, questions have been raised about the nutritional quality of GM crops, possible toxic effects upon their consumers, implications for antibiotic resistance (from the "marker" genes commonly used in GM procedures to check if a bacterium has taken up a new piece of DNA), and possible allergies people might develop to new components of food. So far, there is no real evidence which would give cause for concern on any of these counts – though using antibiotic resistance genes in modified organisms destined for wide dispersal seems unwise and it would be good to see other markers used instead. While inserting a new gene or genes, or silencing an existing one, may give rise to complex effects on an organism's biochemistry, none of the GM crops now planted seem appreciably altered in ways which differ from what the designers intended. Neither nutritional quality nor toxicity are affected.

The allergy problem could certainly arise if someone put a gene for a protein from peanuts into wheat, say, but it is not clear why this would be done. If it was, current labelling of the "may contain nuts" variety would have to be extended!

Environmental concerns include: the potential harm that might be inflicted upon insects and wildlife (either directly or through increased use of herbicides); the spread of resistance to herbicides or pesticides; and the development of **superweeds** that acquire genes from modified crop plants. The larger concern is that unlike genetically modified organisms which remain in the lab, or even in the factory, GM crops are designed to be released into the wider world. Pollen floats on the breeze, and cross-pollination with wild varieties is highly likely, even if GM crops are planted inside a broad, unplanted area to try and isolate them. So there are suggestions that herbicide tolerance would spread – perhaps to wild relatives of the agriculture crop, or even to the weeds themselves.

As with many aspects of the debate, there are two ways to look at this. In an distinct echo of the **recombinant DNA** debate, the transfer of genes between species can be portrayed as unnatural, and likely to have unforeseen consequences – especially if the genes are released into the wild. The understanding of genomes which has developed since the mid-1970s, though, suggests that natural movement of genes, known as **horizontal gene transfer**, is much more common than previously known, and may be an important factor in evolution. This is undoubtedly true in bacteria and may well turn out to be the case, through involvement of genetically unstable viruses, for other organisms too. What should we make of this in relation to artificial gene transfers in crop plants? Well, on one hand it suggests that what people are now doing by design is less new than we thought, and so perhaps less worrying. On the other, it implies that genes will get around in ways we may not expect, so we ought to be cautious which ones we give a helping hand.

The bottom line is that we still have a lot to learn about **gene flow**, interaction between organisms, and the working of ecosystems, so there are many uncertainties about the potential consequences of new combinations of genes. Even if it turns out that they are almost invariably harmless, just one or two with nasty side effects could cause widespread problems. The obvious prescription is more research, careful testing and long-term monitoring. But doing the research requires growing the plants outside the lab in the first place, and – as in the pharmaceutical industry – commercial pressures may compromise commitment to safety tests.

If these are difficult issues to resolve, so are questions about who gets the benefit – including the balance between the interests of producers and consumers, big companies and large-scale farmers, and people in the developed and the developing world.

So there are many specific issues, combined with some sharp differences in outlook. Some of the latter are political, or differences in world-view. Some are more scientific. Some, crucially, are a combination of the two. For example, GM-crop controversies crystallize a difference in approaches to the natural world that became very visible during the debates over **recombinant DNA** (see p.182). Molecular biologists, as the term implies, like to see things in terms of molecules. They tend to think of gene alterations in the context of precise experiments, with well-controlled outcomes. This tempts them to believe that safety will come from thinking through what **gene shifts** to make, and under what conditions.

Their colleagues in ecology, on the other hand, regard such alterations as bringing new uncertainties, which indicate a need for large-scale field testing to find out what will happen when a new combination of genes gets among a complex of other organisms. They worry about the "unknown unknowns".

Add in the operation of a regulatory system which includes a host of national and regional organizations as well as those with a global remit like the World Health Organization or the World Trade Organization, and the field is pretty complex. So there are a range of answers to one question which often arises when people review the GM debate so far: why has the response to GM crops apparently differed in Europe and the US?

Why are US and EU attitudes so different?

The politics of GM foods in the late 1990s were a study in contrasts. In Europe, and especially in the UK, there were banner headlines about "Frankenfoods", consumer boycotts, sabotages of field trials, and supermarkets and restaurants vying to be GM free. In the USA, meanwhile, people happily munched and slurped their way through products with GM ingredients ranging through cookies, peanuts, candy, food tenderizers, aspirin, honey, beer, bread, fruit juice, soft drinks, milk, pickles, cereal, tomatoes and peppers, ketchup and chips, apparently without a care.

The US did see a campaign against use of recombinant **Bovine Growth Hormone** to boost milk production in the early 1990s. There have been controversies in the US in the past about the release of GMOs (Genetically Modified Organisms), and there doubtless will be again. Kraft foods recalled millions of packs of Taco shells in 2000 when environmental groups showed they had been made from a mixture of corn that included an engineered Bt strain only approved for animal feed. This strengthened calls for labelling GM food products in the US, a process which occurs in most European countries. Before this, few Americans were aware of the spread of GM ingredients in processed foods. Nevertheless, it remains the case that many US consumers seem unconcerned about GM foodstuffs, compared with their European counterparts. Even GM developments that met serious opposition in the US, such as **GM wheat**, were mainly fought because farmers feared their crops would not be marketable overseas. As well as commercial use, there have been thousands and thousands of field trials of experimental GM crops in the USA, though these tend to be for testing how the plants grow rather than involving more wide-ranging monitoring of, for example, environmental effects. Yet controversy about biotechnology is not unknown in the USA – think of embryonic stem cells. So the difference over food is intriguing.

Like the question of possible adverse effects of GM crops, the explanation is not going to be simple. Future historians will have plenty of things to take into account. Most of the GM crops are being grown in the US, and it has been easy in Europe to portray GM as an imposition from America (bad) and a tool of big corporations (worse). The British wing of **Greenpeace** tapped into this when they pulled up outside the security gate 10 Downing Street during the Clinton presidency in the USA and dumped a load of GM soybeans from a truck with a banner saying, ahem, "Tony, don't swallow Bill's seed." And they were helped by the recent policies of Monsanto, the leading player in GM seed production, who appeared to be doing everything possible to live up to the image of the big, bad corporation. Monsanto has been involved, as both plaintiff and defendant, in a number of high-profile lawsuits, and has garnered a great deal of controversy over disputes concerning "royalties" payable for the use of its seeds. Needless to say, neither Greenpeace nor Friends of the Earth are exactly fans of the company. For critical assessments of Monsanto-related activities, see www.monsantowatch.org

Differences in public awareness also seem to be part of the story. A survey released in November 2005 by the **Pew Initiative on Food and Technology**, a charitable trust based in Washington DC set up to encourage public debate on biotechnology, revealed that almost six out of ten adults in the US are unaware that GM crops exist, while only 25 percent realize that GM foods have been on sale in the US for the past ten years.

Another big factor was the recent history of British food safety (or lack thereof). The crisis precipitated by outbreaks of "mad cow disease" – BSE (Bovine Spongiform Encephalopathy) – effectively destroyed the credibility of government reassurances based on scientific advice to ministers. And BSE, as well as killing cows and eventually humans, also exposed the realities of modern agriculture to urban consumers who had largely ignored the industrialization of production in return for cheap eats.

More subtly, there is a long-standing difference between the US and Europe in the way biotechnology products were regulated. Put simply, the US regulatory agencies tended to accept the common industrial point of view that the process did not matter – as long as the product was the same as one already in use, the fact that it was made using new-fangled genetic technology was not relevant to safety assessments. European bodies, both nationally and at the level of the European Union, tended to pay more attention to all the steps on the way.

When you consider all the implications the advent of GM food has for consumer choice, the labeling of ingredients, and the rules supposedly governing the propriety of world trade, it is not surprising that it has caused such controversy. In terms of choice, many people in the UK felt they had been caught unawares. The first GM product, canned tomato puree, had been labelled as such in 1996. The subsequent discovery that many processed foods might contain oil from **GM rape** (canola) came as a surprise. You might argue that the oil was just the same as its non-GM neighbours on the supermarket shelf – "substantially equivalent" in the regulators' jargon. But since there were at least some scientists arguing about the risks, and there was no obvious benefit, it seemed prudent to many to just say no. "GM Free", like "organic", was used by stores as a shorthand label for "we know you care what you feed your family, and we care about you", whatever the merits of the actual product.

As we argue below, *blanket* avoidance or unthinking acceptance of GM products, whether the actual crop or some processed product, makes little sense. The way through the thicket of controversy is detailed, case-by-case, assessment. But a shopper filling their trolley in the supermarket aisle has no time for such niceties. For now, GM is on hold in much of Europe not because it is dangerous, or bad for the countryside. There is simply no market for it. A next generation of GM crops, with more obvious attractions to the diner, could well change that. Imagine the appeal, for instance, of vegetables which have been engineered to make the omega-3 fatty acids now often touted as good for the brain. If it came down to a choice between getting one's omega-3 from fish from marine populations in terminal decline, or from omega-3-enhanced soybeans a committed environmentalist would have to think twice. Meantime, the strongest divide between the two sides of the Atlantic seems to be that the citizens of the US mostly trust their regulators, while those of the European Union are more suspicious of their competence to protect health and environment.

GM crops and the developing world

It has been easy for consumers in (some) Western countries to avoid or even to just ignore GM crops. The first direct product of GM – canned GM tomato puree – went on sale in 1996 and sold reasonably well because it was cheaper than non-GM alternatives. But it was eventually withdrawn in 1999, after safety concerns surfaced about another, unrelated GM tomato strain, the so-called **flavr-savr**. That one surely deserved a ban for the name alone, although anti-GM campaigners queried whether enough was known about the possible effects on nutritional quality, production of toxins naturally present in tomatoes, or possible new allergies, which might result from the genetic modification of the tomato to prolong its shelf life. Although flavr-savr was a pretty normal tomato apart from one enzyme which speeds ripening, it too was withdrawn by it makers, Calgene.

But if shoppers in affluent countries can easily avoid GM ingredients, and feel they are not missing any great benefits, could they offer more to areas of the world where many people do not have enough to eat? Food supply is a complicated political and economic issue, of course. Much malnutrition is caused by poverty, not poor supply. But increasing food production is still a worthy goal when global population is expected to go on growing until the middle of the century. Getting farmers in the south to grow GM soybeans to feed cattle for beefburgers won't help there. Altering plants to create hardier varieties which can fight pests or survive drought or salty soil just might.

Then there is the hope of making some food more nutritious by altering genes. Although plenty of processed foods have GM ingredients because farmers produce them that way, when it comes to the actual physical and chemical qualities of specific crops, little has changed since the demise of the Frankentomato. Shoppers cruising the supermarket aisle may yet get excited about coffee from plants modified to grow uncaffeinated beans, or onions which lack the chemical which makes your eyes water when you slice into them, though we rather doubt it. Maybe so, say GM advocates, but think of the benefits to the poor, underfed billions of the south from other, nutritionally more significant changes now being sought. That is where the real potential of genetic modification will be realized. The modified crops already widely used are only a small sample of the new

Pusztai's potatoes

There are plenty of hypothetical ways in which genetic manipulation of crop plants might create a novel hazard. But evidence that they have ever actually done so is scant. So when an established scientist in Britain announced, in 1998, that genetically engineered potatoes were harmful when fed to rats, he got a lot of attention. When he was shortly afterwards suspended, retired by his employer, and roundly criticized by a specially convened review committee set up by the Royal Society, the cries of "cover up" were loud and long.

The controversy centred on plant proteins known as **lectins**, some of which are natural insecticides. **Arpad Pusztai** of the Rowett Research Institute in Aberdeen had been trying to assess the possible effects of GM potatoes, which produce the lectin GNA (no relation to DNA). He fed rats with GNA potatoes, and with unmodified potatoes with GNA added. The results, the Institute said in a series of press releases, showed adverse effects on growth rates and on the rats' immune systems.

It was pretty hard to figure out what the effects were, however, and what might have caused them. Pusztai had not produced a full scientific paper on his findings (one appeared more than a year later). And it was not clear from the press information how he had distinguished between the effects of eating solely potatoes (a poor diet for rats, as it is for humans), of feeding them raw or cooked, and differences in protein content between transgenic and normal potatoes. Nor did he find any differences in growth rates in long-term (110 day) feeding studies using transgenic GNA potatoes and unmodified potatoes without added GNA.

The main finding he drew attention to came from one batch of rats in a short-term (ten day) feeding study which appeared to have more undigested food in their guts. He suggested this meant that "digestion and absorption of transgenic potato diets was retarded in comparison with ordinary potato diets".

strains that could be on offer. There is research under way, most of it publicly funded, in three main areas. Work on improving resistance to viruses and fungal attack has addressed crops such as coffee, bananas, cassava, potato, sweet potato, beans, wheat, papaya, squash and melon. Studies on tweaking genes to make plants hardier – in the face of drought, heat, frost or acid or salty soils – have covered cotton, coffee, rice, wheat, potatoes, brassica, tomato and barley. And there is work being conducted on enriching the vitamin or "micronutrient" content of various species, including rice, wheat, maize, cassava, millet and potato. Boosting the **vitamin A** or **iron content** are the kinds of improvement that are envisaged. The **Danish Institute of Agricultural Sciences** reported in 2006 that it has produced GM wheat plants with a stable version of an enzyme

Others were not convinced he had made a careful enough comparison to support even the suggestion, let alone a firm conclusion.

Intensive reviews followed, first by an "audit committee" set up by Pusztai's employers, who suspended him, then by the Royal Society. Both concluded that the data were weak, and did not support the suggestion of harmful effects from the GM diet. Further details of the immune effects were published in 1999 in the medical journal *The Lancet*. But most of Pusztai's fellow scientists remained unconvinced that they were due, as he argued, to the **promoter sequence** used in the GM potato strain. There were too many other effects to consider, including malnutrition in the rats, and toxins normally found in varying amounts in potatoes. "This study is more informative about working with potatoes than it is about GM technology," said one plant expert quoted in the *New Scientist*.

Some years on, it is harder to see Pusztai as a persecuted whistleblower. A martyr to the truth really needs to start with better experiments. On the other hand, the way he was treated certainly fitted him for the part. In 2002 the *Guardian* reported an unnamed civil servant writing in a memo that "the sight of this heavy-handed scientific community bringing its full academic prejudice to bear on this frail, ageing scientist lost much of the case on sympathy and fairness grounds."

You can find the Royal Society's review, which concluded that "on the basis of information currently available there is no convincing evidence of adverse effects from the GM potatoes in question" at www.royalsoc.ac.uk/document.asp?id=1462

Pusztai's own web page continued to document his exchanges with various other scientists and science organizations, but is no longer updated.

www.freenetpages.co.uk/hp/a.pusztai/

that is vital for release of zinc and iron when we digest the grain. Without the enzyme, phytase, these metals remain bound in a form which makes them unavailable to the diner. The new wheat has a phytase gene from the fungus *Aspergillus fumigatus* which gives it an enzyme that survives boiling, so making bread or porridge does not inactivate it.

More speculatively, there are projects to incorporate proteins from disease-causing organisms into fruits and vegetables, as a cheap route to "vaccination". So eating a banana, say, might protect you against hepatitis – cutting out the need for injections. But there is much work to be done to show that such routes to immunization are effective, that doses can be controlled, and that the plants will cause no other knock-on ecological effects in the wild.

There are so many possibilities for the improvement of crops, however, that the technology of genetic modification is bound to be a part of future strategies for world agriculture. The authoritative, and terrifically reasonable, **Nuffield Council on Bioethics**, an independent British review body, concluded in a 2003 report that weighing the merits of GM crops for developing countries calls for case-by-case assessment, not blanket approval or condemnation. And they stressed that such assessment should focus on the specific situation in a particular country's economy and agricultural system, and compare GM with alternatives.

Some of the existing use of GM does already show benefit in developing countries, even though it is too soon to say if the benefits will last. Their report cites the several million poor cotton farmers who use Bt cotton in China and South Africa, for instance, and finds that the Bt strains have saved money, increased yields, reduced pesticide use and cut health problems caused by spraying chemicals.

Reviewing an impressive list of such cases, the Nuffield Council reckoned that some GM crops have shown they can reduce pressures on the environment and address specific health, ecological and agricultural problems. They add that "in some cases the use of a GM crop variety may well pose fewer risks than the agricultural system already in operation" (see www.nuffieldbioethics.org/go/ourwork/gmcrops/publication_313.html).

However, no such verdict is definitive, as new evidence continually arrives and circumstances change. There are now indications that the early benefits of Bt cotton are being reduced by newly developed insect resistance, for example. This wipes out farmers' profits because they have to resume using large amounts of chemical herbicides and pay for the expensive Bt seed. Arguments about the overall effects are bound to continue. The industry-funded **International Service for the Acquisition of Agri-Biotech Applications** (ISAAA) reported that planting of GM crops in the developing world increased by 21 percent in 2006 – and suggested that this indicates growing acceptance of GM among farmers. **Friends of the Earth** and **Greenpeace** both countered with reports of their own documenting what they said was increased opposition to GM. But both sides were still discussing the first generation of GM crops. The argument will continue as the technology continues to develop.

Patenting life

A patent gives the owner exclusive rights to an invention. This is one way creativity can be turned into money. So a large crew of patent agents, lawyers and officials make a living ensuring that patent specifications describe the invention in enough detail. The patent must allow the invention to be protected, for however long it is granted. And it must describe something genuinely new. A system established to ensure a fair return on nifty widgets – things to make your Spinning Jenny spin faster – is now a vital part of a world of science-based invention and megabuck research budgets. And nowadays it applies to living "inventions" too. This new development coincided with the advent of recombinant DNA technology, and has been just as controversial.

No one has made new life (yet): the first **bio-patents** to be granted were for slightly modified life. In the US, they date from well before the recombinant DNA era, as lobbying from plant breeders secured the passage of the **Plant Patent Act** in 1930. This only applied to plants that were reproduced asexually (clones), as this permitted the tight specification that patents demand. However, the first modern-style patent application on a living organism was granted in 1980. The Supreme Court overruled the **US Patent and Trademark Office**'s decision not to allow an application to patent a genetically modified bacterium that can eat oil spills. The Court ruled that a "live, human-made" microorganism could be patented if it resulted from "human ingenuity and research". That made it a new, man-made object, and thus a "manufacture" or composition of matter within the meaning of the act.

The patent was for a new variant of the *Pseudomonas* bacteria. Ananda Chakrabarty of **General Electric** had fashioned a better oil-eater by combining various plasmids in a single strain. His application, in 1972, came before the explosion of recombinant DNA techniques in the mid-1970s. By the time it was granted, after eight years of legal wrangling, people were modifying organisms all over the place.

The expanding biotech industry, like the drug industry before it, was research intensive and had a keen eye on intellectual property. Before long, patents were as profuse as land claims in a gold rush. Some old-style patents on method – most importantly the application from Stanford University for a patent on Stanley Cohen and Herb Boyer's basic technique for gene splicing in 1974 – were contentious. They would impede research, it was said, they depended on earlier results which had been freely exchanged, and they would see private profit accruing from publicly

Golden Rice

Vitamin A (retinol) is one of those vital chemicals which humans cannot make for themselves and have to get from eating a proper diet. You can eat stuff that has Vitamin A in it, like **cod liver oil** or, more palatably, eggs, meat or milk. Or you can munch on carrots or other fruits and vegetables loaded with **carotenoid compounds**. These are converted, not very efficiently, into the vitamin.

But many millions of people don't have these options: the little they get to eat is mainly white rice, and are thus at risk of Vitamin A deficiency. (Brown rice is OK, but it is harder to keep and often seen as a lower status food.) The WHO reckon that more than 100 million children worldwide are Vitamin A deficient. Between a quarter and half a million of them every year go blind.

One way to reduce this awful toll might be to fix it so that rice plants make **b-carotene**, the best precursor of Vitamin A. The job involved a good deal of gene juggling, as several different enzymes are needed to make b-carotene from its own precursor chemical, and the process took eight years. But in 2000, a team led by the Swiss professor **Ingo Potrykus** announced they had successfully raised a test crop of "Golden Rice", modified to make b-carotene in the edible part of the seed known as the endosperm. Their paper was published in the journal *Science* and quickly made it to the front cover of *Time* magazine.

funded work. Stanford defused criticism by allowing academics to use the method for free, and setting a low license fee for companies. But the same questions arose in other cases (see p.70 to read about the BRCA1 wranglings).

Beyond method, there were soon more patents on organisms. By the late 1980s, genetically modified plants and animals were being patented. **Philip Leder**'s group at Harvard created an "**oncomouse**", which develops breast cancer, by inserting an **oncogene** (a cancer-causing gene) and a promoter into a fertilized mouse egg. His sponsor, **DuPont** claimed rights to the method and secured a blanket patent on transgenic animals that have been modified to develop cancers.

There was also extensive patenting of genes themselves (including human genes). These could be artificial, or even just newly isolated and

Could this be the "killer app" for GM crop technology? The work was publicly funded and the sponsors planned to make the rice available for free to farmers. Surely this was science delivering something good? But critics were not convinced. They argued that Vitamin A deficiencies were largely due to the earlier "Green Revolution", which had encouraged cultivation of high-yielding rice strains at the expense of traditional mixed farming. The answer was to reintroduce mixed cropping and to include existing species that had been used in the past. Besides, nobody could eat enough Golden Rice Mark I to meet their daily vitamin requirement. The b-carotene content was low, and rice lacked the fat and iron that promote its absorption and conversion. The fact that the new rice grain was yellow didn't help, as it was thought likely to deter people in many cultures from eating it. A better solution might be to promote a more mixed diet using traditional plants already found in rice-dependent areas.

The promoters of the new grain are undeterred. They accept that a varied diet is best for everyone, and that simple technical fixes are not always a good fit with local custom. But, they say, these are poor reasons to reject a possible solution when the ideal one is unlikely to be realizable quickly. They are working to improve the properties of the rice until it can provide the recommended daily intake of vitamin in a 100–200 grammes portion, about as much as a child in rice-based societies may hope to eat.

Golden Rice may also be developed further into a plant which has other added benefits from new genes. In June 2005 the Grand Challenges in Human Health Project, an initiative launched by the Bill and Melinda Gates Foundation, announced support for a German university team trying to develop rice which had better quality protein, high vitamin E and enhanced iron and zinc content, as well as making lots of b-carotene.

You can read more at www.goldenrice.org

defined. Between 1981 and 1985, over 1000 human gene patents were granted worldwide, and there are more than 40,000 DNA-based patents. As this is quite a lot more than the number of individual human genes, you get the impression that the commercial world has the human genome pretty well staked out.

The idea of patenting genes (or in some cases **cell lines** – colonies of cells derived from a single modified founder) has continued to generate controversy. Whose genes are they, anyway? If a group of gene prospectors takes medicinal plants home from the rainforest, do the tribespeople, who already recognized and used the plant, have any rights in the gene patents that follow? If a man's pancreas is surgically removed, should a cell line established from his tissue belong to the researchers?

Bringing DNA into the patent system has been criticized on several different counts, usually by NGOs. One is that DNA, especially human DNA, should have special status, and should be regarded as part of a common heritage, not something which can be owned. Then there are more technical arguments about whether DNA sequences alone meet the legal criteria for patenting. And there are concerns about the implications for health care and research if patents restrict access to some information.

Some of these arguments can be left to the experts. The idea that DNA should not be patentable because it evolved naturally is more interesting, however. Constructing new combinations of genes may qualify as invention, as the court ruled in the Chakrabarty case. But just fishing out pre-existing DNA sequences is surely discovering something which already existed. The additional objection to patenting human DNA seems to derive from a more widely accepted restriction on property rights. Owning humans is illegal – we call it slavery. And most patent regimes extend the right to own one's own body to *parts* of one's body, including bits of its DNA. As a key EU directive puts it: "The human body, at the various stages of its formation and development, and the simple discovery of one of its elements, including the sequence or partial sequence of a gene, cannot constitute patentable inventions."

Can you patent yourself?

In February 2000 a British woman, **Donna Maclean** applied to the Patent Office in London for a patent on herself. She claimed she was both novel and useful. "It has taken thirty years of hard labour for me to discover and invent myself, and now I wish to protect my invention from unauthorized exploitation, genetic or otherwise," she wrote in her application. "I am new: I have led a private existence and I have not made the invention of myself public. I am not obvious."

Ms Maclean said she had many industrial applications. "For example, my genes can be used in medical research to extremely profitable ends," she wrote. "I therefore wish to have sole control of my own genetic material." She got a polite reply from the Patent Office assigning her application a number but saying it could not be processed further without payment of their £130 fee.

Not to be outdone, San Francisco artist **Marilyn Donahue** outlined an imaginative procedure for copyrighting DNA. Have someone take a snapshot of you licking a first-class stamp. Then put the stamp on a self-addressed envelope containing the picture. The two together authenticate the saliva sample. Unfortunately, the US Copyright Office did not recognize her right to creative control over her DNA.

This sounds straightforward. But the same document says the sequence *can* be protected by a patent if it is isolated from the body, "even if the structure of that element is identical to that of a natural element". As Britain's **Nuffield Council on Bioethics** comments, it is hard to understand how these two principles can be consistently combined. Maybe that is why lawyers get paid so much.

But the biggest dispute so far centred on the biggest project, the effort to map and sequence the **human genome**. The rival teams, one publicly funded and one privately funded had diametrically opposed views on the release of the information (see p.77). The private consortium essentially sought to secure ownership of the data, and grant access to it on subscription. Patenting was talked about, though never quite followed through. At one stage in the conflict, the British **Wellcome Trust** charity got together with ten big drug companies to establish a foundation to find and map **single nucleotide polymorphisms** (SNPs, see p.97). Their plan was to publish them all, but only to prevent anyone else (ie **Celera**) patenting the same information.

The human genome is now public property, an outcome confirmed by Celera's announcement in 2004 that it was depositing its human genome sequence in the online database **GenBank**. But whole **genome patenting** is a reality in other organisms, albeit so far only in viruses and bacteria.

Meanwhile, the growth of **stem-cell research** has led to a new round of patent conundrums, as **cell lines** with different properties arise and their developers seek patent protection. The UK Patent Office, for instance, ruled in 2003 that human embryonic stem cells which are "pluripotent" – that is, ones that can give rise to a range of different tissues – can be patented. However, the official statement of the latest position went on to say that: "Human totipotent cells have the potential to develop into an entire human body. In view of this potential, such cells are not patentable because the human body at the various stages of its formation and development is excluded from patentability by Paragraph 3(a) of Schedule A2 to the Patents Act 1977. The Patent Office will therefore not grant patents for human totipotent cells." So, reassuringly, it looks as though human beings are still patent free – for now.

Cloning

Over the last thirty years, the ability to isolate, study and manipulate individual genes has spawned whole new fields of study, and a host of practical applications. Yet mention the word "genetics" to many people and their first thought will not be DNA fingerprinting, medical diagnostic tests or human insulin – it will be **animal cloning**. Geneticists are often slightly baffled by this, since cloning technology has much more in common with reproductive biology than anything done in a genetics lab. Cloning involves manipulating the entire genetic information of an organism, while molecular genetics involves the manipulation of single genes or pieces of DNA.

It's true that there may be an increasing overlap between the two areas, particularly in "**pharming**" (the creation of animals genetically modified to produce human proteins – see chapter 10), and perhaps in the development of "tailor-made" **stem-cell** therapies (see chapter 13). But to date, cloning has remained the preserve of a few agricultural research and developmental biology labs. In this chapter, we'll look at how cloning works, why anyone would want to do it, and why **Dolly** has become the "poster animal" of modern biology.

Cloning animals

Unlike extracting DNA from onions (see chapter 3), cloning an animal is not really something you can do in your kitchen. Cloning **plants** is another matter – people have been doing it for thousands of years. If you've ever taken a leaf cutting from a plant and grown it into a new one, then you've created a clone: an organism with the same genetic information as an existing one. A more sophisticated version of this process involves isolating single cells from the root tips of a plant and growing them in the laboratory. This method, called **tissue-cell propagation**, can be used in horticulture and agriculture to produce thousands of identical plants, and is also the key to genetically modifying a plant (see chapter 3).

Cloning animals is harder, because with a few notable exceptions, most animals reproduce sexually – that is, they need genetic information from both a male and female. While yeast and some creatures like hydra can produce offspring simply by "budding" – forming an outgrowth that drops off and forms a new organism – the same is not true of sheep. So animal cloning in the laboratory involves taking the genetic information from a single cell of an existing animal, inserting it into an unfertilized egg emptied of its own genes, and somehow persuading the embryo to start developing.

The full name for this technique is **somatic cell nuclear transfer** (SCNT) – a term that scientists have been trying to get journalists to use in place of the word "cloning" for many years, without much success. Although Dolly's birth was announced in 1996, SCNT has in fact been around for much longer. In 1952, US scientists **Robert Briggs** and **Thomas King** carried out the first successful SCNT experiments, in which they managed to clone the embryos of a species of leopard frog (*Rana pipiens*). They took a single cell from a frog embryo in the early stages of development, and, using a glass pipette (a fine tube used for sucking up liquids), removed its nucleus – the structure at the centre of a cell that contains nearly all of its genetic information. They then inserted it into an enucleated egg cell, one from which the nucleus had been removed. The resulting cells sometimes grew into whole new embryos and, in later experiments, tadpoles. Once honed, their technique had an impressive success rate of about forty percent.

Despite their success, Briggs and King were convinced that SCNT would only work using nuclei from early embryos, and not cells taken from adult tissue, which are fully "**differentiated**" – that is, committed to a being a particular type of cell. A fundamental question of the time was: do the cells of an adult organism specialize by losing all their genes, except the ones they need to carry out their specific function (eg to be a liver or skin cell), or do they simply switch off the genes they don't need? This problem had originally been posed by German embryologist **Hans Spemann**, who carried out the experimental forerunner of SCNT in 1938. He used a loop of hair to constrict a newly fertilized **salamander egg**, so that the nucleus was in one side and the rest of the cell's contents were in the other. After the side with the nucleus had divided four times, creating an embryo made up of sixteen cells, Spemann loosened the hair and allowed one of the nuclei back over to the enucleated side. It started to divide too, and by tightening the hair, Spemann was able to split the embryo into two, each of which developed into a salamander. This work showed that each cell in a very early embryo is still "totipotent", that is, able to develop into any type of body tissue.

Twins: natural born clones

Human clones have been walking amongst us ever since our species evolved. Identical twins usually arise if the early embryo splits into two any time before or during the so-called **blastocyst** stage of development. Depending on exactly when the split happens, twins may share a placenta, or even an amniotic sac. Identical twins don't run in families – they happen around once in every 250 births, for reasons unknown. Very occasionally, either through incomplete separation or subsequent fusion of two early embryos, identical twins can be conjoined. Non-identical twins are formed when two different eggs are fertilized by two different sperm, and so are no more genetically alike than ordinary siblings. Embryo splitting has been used in agriculture and animal cloning research for several years, to artificially create genetically identical individuals. But the success rate is not very high, and there is a limit to the number of "cloned" embryos that can be produced in this way.

Briggs and King's belief that the "genetic potential" of a cell greatly reduces as it specializes was challenged by work published in 1958, by UK scientist John Gurdon. Using a different species of frog, *Xenopus laevis*, he managed to clone tadpoles using the nuclei of fully differentiated cells, taken from the intestinal lining of adult frogs. Although none of the cloned tadpoles grew into frogs themselves, Gurdon's work showed that nuclei from fully specialized cells still contained all the genetic information necessary to make a complete new animal.

Many scientists were initially sceptical of the results, because of Briggs and King's marked lack of success. But further experiments by Gurdon, in which he created adult cloned frogs using eggs with genetically "tagged" nuclei taken from **tadpole cells**, proved conclusively that cell specialization is a reversible process – genes stay the same, it's the switches that make the difference (so-called **epigenetic changes**, see p.135).

However, Gurdon's success rate using adult cell nuclei was nothing like as high as that achieved by Briggs and King using nuclei from early embryos, showing that "reprogramming" a cell's genetic information is increasingly difficult to do the further it has advanced down a developmental pathway. A string of failures in other species also persuaded some that the technique would never work in mammals.

But a sheep proved them wrong. To clone Dolly, **Ian Wilmut** and **Keith Campbell**, of the Roslin Institute in Scotland, used a nucleus taken from the udder of an adult Finn Dorset sheep. (Dolly's name is a nod to well-endowed US country singer **Dolly Parton**.) They put the nucleus into an enucleated egg from another breed of sheep, the Scottish Blackface, and applied a brief jolt of electricity to get the cell to start developing into an embryo. This was then transferred back to the womb of a Blackface ewe, and allowed to grow into a lamb.

Though eventually successful, Dolly's birth followed 276 failed attempts. The Roslin Institute team showed for the first time that it is possible to clone an adult mammal using SCNT, but their efforts also underlined the difficulties of the procedure. Since Dolly's birth, mice, rats, cattle, pigs, goats, horses, rabbits, mules, cats and a dog called **Snuppy** have joined her in the cloning hall of fame.

Problems with cloning

Dolly was born on July 5, 1996 and was put down at the age of six in 2003, suffering from a lung disease more common in older sheep, though not before she had had healthy lambs of her own. When she died she was also obese, and affected by arthritis. Her stuffed remains are on display at Edinburgh's Royal Museum. The jury is still out on whether her untimely death was due to the unusual nature of her creation, or her pampered celebrity lifestyle.

But attempts to clone other animals certainly back up the idea that cloning isn't the most healthy way to start life. Cloned animals – the few that make it past the foetal stage – have been affected variously by enlarged organs, and problems with their immune system, joints and liver, as well as obesity.

Some of these conditions are thought to be down to errors in the reprogramming of genes during the **SCNT procedure**, particularly to faulty "imprinting" – the process by which certain mammalian genes are switched off during early embryo development, according to whether they were inherited from the father or mother. For some imprinted genes, only the version inherited from the mother is switched on, while the paternal version is silenced. Others show the opposite pattern – only the paternal version is switched on, while the version inherited from the mother is switched off. The reprogramming process, in which genes are "tagged" to show that they come from either the mother or father, happens while the egg and sperm are maturing. It's thought that creating embryos using SCNT may mess up this genetic fine-tuning, since it bypasses egg and sperm production altogether.

Another possible reason for the ill health of clones is that their "**cellular age**" can be older than their biological age. The age of a cell can be inferred from the length of its **telomeres** – protective pieces of DNA at the ends of chromosomes, which shorten every time a cell divides and copies its genetic information. So telomeres act as a kind of **cellular clock**, marking the number of times a cell has divided to make two new cells. When telomeres get very short, the cell has difficulty working properly, and it eventually dies. Dolly, like many cloned animals, had shorter telomeres than other sheep her age, which may have caused her to age prematurely. But this isn't true of all clones, or even *clones* of clones: in 2000,

Ian Wilmut

British researcher Ian Wilmut is known to the world's media as the "father of Dolly", the leader of the team that produced cloned sheep, and the indefatigable defender of the motives of the experimental cloners. Born and raised in the English Midlands and Yorkshire, Wilmut was drawn to farms and the outdoors as well as science. He was set for a quiet career as a dogged, rank-and-file agricultural researcher – his doctoral thesis was on freezing boar semen and his first post-doctoral fellowship came from the Milk Marketing Board.

But he had a first taste of media fame when he led efforts to produce a calf from a frozen embryo transplanted into a surrogate cow mother. The year after that calf, "Frostie", was born, in 1974, he moved to the Scottish research laboratory which became the Roslin Institute outside Edinburgh. There, he continued to work on embryos, but switched to genetic modification in 1982, at the behest of the lab's new director who wanted to modernize its programme.

The cloning work was a team effort, and depended heavily on the insights of Wilmut's colleague **Keith Campbell**, who arrived at the Institute in 1991. He was interested in the fundamentals of cell biology and development, and had the crucial insight that successful nuclear transfer depended on getting the cell cycles of donor and recipient in step. This led to the successful conception of two cloned sheep, **Megan** and **Morag**, from nuclei taken from embryo-derived cells, in 1995. Dolly followed soon after.

The huge publicity which ensued also exposed some of the tensions which existed there, as in many scientific teams. Although Campbell co-authored a popular book about Dolly with Wilmut, rumours that he felt he had not been given his due persisted. Wilmut's later assertion (in an unrelated industrial tribunal hearing) that his co-worker should have two thirds of the credit, himself one third, served mainly to annoy his technicians, who had slogged through hundreds of nuclear transfers and embryo implantations. Wilmut had to delegate all this work as he has a hand tremor which means he cannot do the delicate manipulation of cells involved. So they felt they deserved co-authorship, rather than the footnote acknowledging their contribution that appeared in the technical paper outlining how Dolly was born.

On the other hand, the reassuringly ordinary, schoolmasterish Wilmut shouldered a hefty burden of interviews, hearings and lectures in which he explained cloning to the world, defended its uses, and argued consistently for the potential agricultural and medical benefits, while opposing any effort to clone a human.

a Hawaiian team found that six generations of sequentially cloned mice aged normally (the experiment came to an abrupt end when the sixth generation clone was eaten by its foster mother). Whatever the reason for the technical difficulties, the problems and low success rate associated with cloning have led to *human* cloning being outlawed in many countries (see box on p.213).

The time he devoted to this undoubtedly impeded his own work. But his group at the Roslin Institute is still pursuing the possibilities of cloning and stem-cell biology. In 2005, he was granted a license by the **Human Fertilisation and Embryology Authority** for experiments intended to use cloned human embryos. Like other embryo research in Britain, these would only be permitted on embryos fourteen days old or less. They were designed to recover embryonic cells which would shed light on the origins of **motor neurone disease**.

Keith Campbell, meanwhile, had moved in 1999 to become a professor at Nottingham University, where he also continues to work on embryonic development and ways of producing stem cells. Wilmut now holds a chair at Edinburgh University, where he leads the University's brand new Centre for Regenerative Medicine.

He has produced two books outlining the origins of Dolly and the uses of cloning: *The Second Creation* (2000), with Colin Tudge and Campbell, and *After Dolly: The Uses And Misuses Of Human Cloning* (2006), with Roger Highfield.

Why clone anyway?

Given the problems associated with animal cloning, why would anyone want to do it at all? Dolly's birth was part of an ongoing research effort to find ways of mass-breeding animals with useful qualities. In this respect, the most important sheep to come from the Roslin laboratory was not Dolly at all, but the lesser-known **Polly**. As we have already mentioned (see chapter 3), Polly is at the forefront of "pharming" research – the use of genetically altered plants and animals to produce pharmaceutical products such as human medicines. Polly was cloned using a nucleus from a foetal sheep cell, grown in the laboratory, to which had been added a gene that makes a human clotting protein called factor IX. This protein is missing in people who have haemophilia type B, an inherited bleeding disorder. It's hoped that factor IX from the milk of sheep like Polly will pave the way for cheaper, safer treatments for **haemophilia**. And using cloning technology means that the tricky genetic alteration bit can be carried out on cells growing in the laboratory, which is much easier than trying to breed individual genetically altered animals.

Apart from making human medicines, cloning may help boost the numbers of some endangered animal species. In January 2001, **Noah**, the first cloned **gaur** (a species of wild ox) was born, although unfortunately he died of an infection just two days after his birth. Noah was created using the nucleus of a frozen skin cell taken from an adult male gaur that had died eight years previously, which was then fused with an enucleated egg from a domestic cow. Also in 2001, scientists in Italy successfully cloned a healthy baby **mouflon**, an endangered wild sheep, while 2003 and 2004 saw the arrival of a cloned **banteng** (a rare Javanese cattle) and a cloned African wildcat. Cloning extinct, rather than endangered animals presents a much greater challenge to scientists, because the egg needed to create the cloned embryo has to be from a different species. And cloning animals that have been gone for millions, or even mere thousands of years, will remain the stuff of science fiction. SCNT requires an intact cell nucleus, rather than the scraps of DNA to be found in ancient bones.

Finally, animal cloning may offer some solace to people who have lost a beloved pet. The first owner of a commercially cloned pet was an airline worker from Dallas, who paid $50,000 to a Californian company to clone her pet cat Nicky, who had died aged 17. **Little Nicky** was born in October 2004, and is said by her owner to have exactly the same personality as the original. But this is unlikely to be true of all cloned animals, since personality is influenced by far more factors than genes. Since coat markings

Human cloning claims and laws

Scientists have not yet managed to clone a primate, despite concerted efforts to do so. In 2003, US researchers reported using **SCNT** to create 716 cloned rhesus monkey embryos, but all of them failed to develop. It seems that there may be something fundamentally different about primate embryos that means they will never be amenable to cloning methods that work for other mammals. For this reason, any claims of cloned humans should be treated with the utmost scepticism. But there have been some, and not all made by religious cult leaders. In 2001, fertility doctors **Severino Antinori** and **Panayiotis Zavos** announced plans to use SCNT to treat ten infertile couples. Antinori had already hit the headlines in 1994, when he helped 63-year-old Rosanna Della Corte give birth to a son, making her the world's oldest mother at that time. Antinori and Zavos later claimed that some of the women they had treated were pregnant, but the cloned babies never materialized. Then, at the end of 2002, Clonaid – an offshoot of the Canada-based **Raelian cult** that believes life on Earth was started by aliens cloning humans – announced it had achieved the birth of the first human cloned baby. "Baby Eve" timed her holiday season arrival well. Allegedly born on December 26, when competition for the headlines was likely to be minimal, the mysterious baby girl received the full attention of the media. Clonaid, which also claimed to have created four other clones, never produced any proof that the baby was a genetic replica of her mother, or offered any details of its methods. Even Antinori expressed doubts that Clonaid had succeeded, and the announcement was widely viewed as a bizarre (and highly effective) publicity stunt.

Despite their unreliability, human clone claims sparked immediate interest from legislators worldwide. Almost immediately after Antinori and Zavos' first announcement, hearings opened in the **US Congress** about the feasibility of human reproductive cloning and a potential ban on the practice. Testimonies from scientists warned that cloning procedures were not safe enough to use on humans, and would result in a high rate of miscarriage and deformities. In August 2001, after just six hours of debate, Congress passed a bill banning all forms of human cloning (both for reproductive and research purposes). In the UK, the Government also swiftly passed a one-clause bill, the **Human Reproductive Cloning Act 2001**, which criminalized attempts to clone human beings – but not so-called **therapeutic cloning**, in which human embryos are created using SCNT for medical research. In 2001, all the United Nations member countries agreed in principle on an international treaty banning reproductive cloning. But before it came to a vote, an alternative, US-backed proposal was put forward, which sought to ban both reproductive and research cloning. The arguments surrounding therapeutic cloning, like any form of embryo research, touch on highly contentious ideas about when life begins. Not surprisingly, the new proposal split the UN, and it has so far failed to agree on any form of human cloning treaty, opting instead for a non-binding declaration. Apparently oblivious to all this worldwide condemnation, in 2002 Antinori again claimed that women in his care were pregnant with clones, but again offered no proof. At this point, Zavos severed all ties with his former colleague, and little more has been heard from the "cowboy cloners" ever since.

are also hard to guarantee (see box on p.215), cloning a cat is something of a gamble. Also, a cloned animal will never be completely identical to the original animal, since as well as the genetic information present in the cell nucleus, a small amount is located in the mitochondria – the tiny, sausage-shaped powerhouses of the cell (see chapter 1). The mitochondria of a cloned animal will always come from the donor of the enucleated egg, rather than the animal from which the cell nucleus originated.

Strange new hybrids: where fact and fiction meet

The tenth anniversary in 2006 of the birth of Dolly the cloned sheep prompted reflections about the enormous media and political response to her arrival. A formerly obscure animal research institute just outside Edinburgh suddenly became world famous, and the hapless sheep was besieged by TV crews and photographers. Why all the fuss?

Certainly, her delivery was a scientific achievement. Dolly developed from an embryo constructed from an enucleated egg injected with a nucleus from cultured cells taken from an adult animal. Making that work was technically tricky, and involved hundreds of failed experiments. It was not, however, the dawn of a new era. The birth of the first "test-tube" baby, **Louise Brown** in 1978, which caused a similar furore, has led to over three million more children around the world born through *in vitro* fertilization. Dolly has led to a few more cloned animals. She also, indirectly, caused much huffing and puffing in parliaments and official committees, and an outpouring of learned assessments from **bioethicists**. But all this had more to do with the fascination of cloning than with scientific reality. To see why, consider what came before Dolly, in science and in fiction.

In science, the role of eggs and sperm in passing genetic information on to the next generation was long an established fact of biology. But as we have said, there were doubts whether the specialized cells of an adult organism still contain all the genetic information present in the newly-formed **zygote**, from which they all descended. Why would they keep reproducing information they no longer needed? Then, as results from frogs, mice and, later on, cows and sheep slowly accumulated, it became apparent that the genes in cells long since adapted to function as part of, say, a liver or a piece of intestine could nevertheless be induced to stir themselves and orchestrate the development of a new egg.

There was still much for **Ian Wilmut** and his colleagues at Roslin to do. Crucially, they had to manipulate the cell cycle of the donor cells so that an egg with its genes removed would accept the replacement genetic package inside a newly injected nucleus. But their success rested more on technical brilliance than on any startling conceptual advance.

The first mammal to be cloned was always going to elicit the biggest response from the word at large. Cloning, or propagation, in plants, was no great threat to anyone. Cloning DNA in bacteria had triggered its own response (see p.182 for the story of the recombinant DNA debate), but the details were hard to follow. The cloning of frogs was startling, but still not all that close to home. But Dolly was almost one of us, it seemed. The pic-

The mysterious case of the cloned cat

Although the world's first cloned cat, **CC** (short for "Carbon Copy" – or "Copy Cat") was a scientific breakthrough, she didn't look much like the original genetic donor, **Rainbow**. This is because some of the genes for coat colour are located on the X-chromosome, of which female cats have two copies, while males have one X and one Y. Since only one active X-chromosome is required in a cell, one of each pair in every cell of the body is randomly – and permanently – switched off. This phenomenon is called **X-inactivation**, and occurs early on in embryo development. So if there are different versions of coat colour genes on the two X-chromosomes, the result is patches of different colours over the animal's body. Rainbow had patterns of black and orange fur on her body, growing from cells that had chromosomes with either the orange or black coat-colour gene

inactivated respectively. CC has no orange in her coat, which shows that a) the cell nucleus from which she was cloned had an active X-chromosome with a black coat gene and an inactive one with an orange coat gene, and b) that the cloning process does not "reset" the inactivation process, as would happen in normal reproduction. Both cats also have large patches of white fur, but they are controlled by different genes altogether.

It appears the appeal of cat cloning, at current prices anyway, is not widespread. The US company **Genetic Savings and Clone**, which cloned two cats in 2004, went out of business in October 2006, selling off its tissue bank to livestock biotech firm ViaGen.

"Carbon Copy" – the world's first cloned feline

tures hardly created an image of a killer sheep about to take over the world (though the American supermarket tabloid the *Weekly World News* did its best to portray Dolly as a red-eyed monster who killed and ate lambs). But if a sheep could be cloned, could a human be far behind?

The question tapped into two sets of ideas and fears which surround cloning. For one thing, ever since the very beginnings of Western culture, there have been myths and stories that feature doubles, twins and doppelgängers. They tend to be a source of confusion, at best, and more often present a threat. A twin can usurp your throne, or sleep with your lover. Something usually ends up being stolen, and it is usually bound up with the sense of identity. **Shakespeare**'s *The Comedy of Errors*, based on ancient Greek and Roman tales, exploits the confusions resulting from two sets of identical twins mistaken for each other. Underlying the humour in this still popular play is the uneasiness that the doppelgänger provokes.

Then there are stories that are more specifically about recent biological developments, real or imagined. The classic example, of course, is the **Aldous Huxley** novel *Brave New World*. Huxley's vision of a dystopian society has long been cited as a warning about mass production through cloning. The significance of artificial creation married to mass production has been talked up many times since Huxley described "Bokanovsky's process" in 1932. It was a technique for multiplying embryos destined to be grown into babies in the lower categories of Huxley's highly regimented future world – the gammas, deltas and epsilons – but is a dead ringer for what we now call cloning. There are even numerous examples in non-fictional commentaries on biology, at least since the late 1950s. The French biologist **Jean Rostand**, for example, in his late 1950s book titled *Can Man Be Modified?* discussed human clones at length.

Rostand's little book described Briggs and King's early experiments with nuclear transplantation (see p.207) in frogs' eggs. Rostand proclaimed that "this new technique of generation would in theory enable us to create as many identical individuals as might be desired. A living creature would be printed in hundreds, in thousands of copies, *all of them real twins*. This would, in short, be *human propagation by cuttings*." From then on, human cloning became a staple of accounts of the biological wonders – or horrors – to come. It featured in more widely read tomes such as the British journalist **Gordon Rattray Taylor**'s *The Biological Time Bomb*, from the late 1960s, which was largely a rehash of Rostand. It came up in countless other books and newspaper articles, often illustrated with multiple images of Mozart or Hitler. The new biology might involve lots of tinkering with hormones, messing with sperm and eggs in Petri dishes, and doing unspeakably complex things with molecules no one could see. But like the "test-tube" baby, cloning was an idea everyone could easily get hold of, even if it was not actually happening.

Or was it? The temperature rose another notch when **David Rorvik** published his 1978 book *In His Image* – which purported to be a non-fictional account of the first human clone. Rorvik, a US science journalist, claimed he had assisted in the cloning of a human being. He presented his story as fact, and provided extensive references to back up his claim that it was possible. This was strongly denied by scientists, most notably by the British biologist **Derek Bromhall**, angry that his own work had been cited. Bromhall also accused Rorvik of presenting fiction as fact out of greed, though Rorvik himself wrote an odd little epilogue where he hoped that "many readers will be persuaded of the possibility, perhaps even the probability of what I have described and benefit by this 'preview' of an astonishing development whose time ... has apparently not yet quite come."

So Rorvik seemed to be admitting that the book was trying to exploit the difficulties scientists have in responding to claims like his. By blurring the line between fact and fiction, he hoped to provoke public discussion. It seemed to work. The book was widely condemned as a literary confidence trick, but it was also widely read. What was really a badly written novel with copious discussion of bioethics, became a Literary Guild selection, and the US paperback rights were sold for a quarter of a million dollars.

All of this fed in to reporting of genuine science, and scientists simulta-neously deplored it and played up to it. A few years before Dolly made her debut, for example, cloning made the front page of the *New York Times* in 1993, and leapt from there to the covers of *Time* and *Newsweek* among an enormous volume of other comment and condemnation. **Gerry Hall** and **Robert Stillman** of the George Washington University Medical School in Washington achieved embryo splitting and announced it as "cloning". They claimed to be amazed by the strength of the global response. The treatment in the news magazines was pretty much identical to the cov-erage of Dolly. Journalists reported the science, but also rehearsed the fiction of cloning. Along with mentions of Rorvik, cloned dinosaurs and legions of cloned Adolf Hitlers, came the suggestion from *Time* magazine that "when it comes to dealing with cloning, ethicists and science-fiction writers have almost identical job descriptions". We take up some of the other fictions later on, in chapter 17. Here, we move on to the arguments from the bioethicists.

Arguments for and against human cloning

At the moment, human reproductive cloning is banned in most countries that have any policy at all on the matter. But if cloning ever became a reliable technology, are there any points in its favour? And who, apart from dictators and the madmen of Hollywood B-movies, would want it?

To start with, it is increasingly considered that any *comprehensive* ban on cloning restricts freedom – reproductive freedom, specifically. This is a newly defined right, as human rights go, but it is mostly upheld (save for religious objections in certain cases) when it comes to choice around contraception and technologies for overcoming infertility. So there may be reasons to forbid cloning, but they might need to invoke other rights that inhibit reproductive freedom, or serious harms attached to cloning but not to other new-fangled ways of reproducing.

But who might want to take up the option of cloning? Well, it would allow some infertile couples, those few who have to use *in vitro* fertilization with donor sperm *and* eggs, to have children who were biologically related to one parent. Similarly, it might make it easier for same-sex couples, or sole parents, to have children who shared their genes.

Cloning could allow some couples to have children without risking passing on a genetic disorder carried by one parent. And it might appeal to some who hoped to have a new child who would be a compatible transplant donor for an older sibling who is critically ill. This one sounds alarmingly close to dramatic scenarios featuring farmed clones as organ donors. But it is little different, in principle, from "choosing" embryos for IVF for similar reasons, which is now allowed for some conditions.

Finally, an individual benefit unique to cloning is that it would theoretically allow parents to have another child with the same genetic make-up as one who had died prematurely. The fact that such a child could never be a replacement, and the parents might do better to just have another child the old-fashioned way, does not necessarily mean that they could not take satisfaction from a cloned baby in the face of their loss.

On top of these there are possible social benefits of cloning, like perpetuating the genetic qualities of outstanding people, or just helping research on human development. But these seem dubious, scary, or just not very impressive, compared with the individual benefits which might be on offer. So if some principle of reproductive freedom suggests people ought to be allowed to try and get these benefits, if they want, what is the case against?

There are safety issues, of course, and problems with consent, so there is a view that just attempting human cloning amounts to unethical experimentation on the unborn. Bear in mind, though, that just the same could be said (and was) about *in vitro* fertilization before the birth of **Louise Brown**. The team who helped her into the world judged that helping her parents to have a baby justified the risks. Others disagreed, but perhaps a lot more animal experiments could have won them over. So let us assume that cloning can be made, if not safe, then about as hazardous as other forms of reproduction we have got used to. When we've seen a few thousand cloned cats, things may look different.

That leaves arguments which emphasize the unusual nature of a cloned person. That is clearly the source of the gut feeling that cloning would be a bad thing which drives much of the politics around the issue. **Leon Kass**, the first chair of President George Bush's bioethics committee, has long argued that this kind of reaction says something important about new biological technologies. We should heed, he says, "the wisdom of repugnance".

That does not seem a very satisfying argument. If the gut feeling is about identity, however, perhaps it highlights some other ideas about rights which matter. One could be a right to a unique identity. If this is a human right, it is presumably waived for identical twins (though the philosophers would say a human right can only be violated by other humans' action, not by a natural event). But if such a right is worth having, cloning only threatens it if we assume that having identical (nuclear) genes makes people identical – and there are plenty of good reasons why that isn't so. As with the furore over GM foods, part of the anxiety over cloning surely arises because many scientists have made such a loud noise about the importance of genes.

Another possible right that cloning might compromise is the right to have an open future. A clone would, it has been argued, know too much about how they were going to turn out for their own good (if they believed in the importance of genes, that is). Even if they defined themselves in opposition to their older – or perhaps deceased – twin, their gene donor would play a huge, and altogether unwholesome, part in their life. Even without a right to ignorance, there would inevitably be some degree of psychological distress involved. So this argument shades into the commonly felt intuition that clones would be liable to turn into screwed-up individuals.

There are other arguments. Some seem far fetched. Yes, cloning would reduce genetic diversity, but this would only matter if a *lot* of people did it. That seems highly unlikely. As sociologist **Alan Ryan** asks, "why would we wish to produce a dubious copy of one parent when we can produce a more interesting non-copy of two, and by so much more agreeable a process?" Other arguments against cloning apply generally to many reproductive interventions. Perhaps the whole set of technologies, from IVF onwards, is moving us toward seeing people as commodities, as means rather than ends, and is disrupting ancient patterns of human relationships that we disturb at our peril. These are all worth pondering, but do not seem to present that much more of a problem with cloning than with the other procedures.

So are we just against cloning because we don't like it, and is all the rest merely blather? It sometimes seems so. It is hard to read many of the systematic discussions of the points for and against without feeling that the conclusion was decided in advance and that the arguments were lined up with that in mind.

Nevertheless, it is not a simple issue. One of the best treatments of the arguments for and against is still an early post-Dolly essay by the subtle US bioethicist **Dan Brock**. You can find the full essay in Martha Nussbaum and Cass Sunstein's collection *Clones And Clones*. Some of the others derive from papers published by the US President's Council on Bioethics, whose predecessor, the National Bioethics Advisory Commission published Brock's essay. The Council was appointed by President Clinton, and was insufficiently conservative in outlook for his successor, **George Bush**, who dissolved it and created the new Commission. You can find the President's Council 2002 report on *Human Cloning And Human Dignity* at www.bioethics.gov/reports/cloningreport/index.html

DNA detective work

The digital identifier

DNA fingerprinting, aka **barcoding** or **profiling** must be the best-known technology making use of DNA. It may not have got a result for the prosecution in the **O.J. Simpson** trial but the folks on *CSI* use it all the time. And there are frequent press reports of real criminal cases that have been cracked using the suspect's DNA. It is not, strictly speaking, a technology directly concerned with genes, as originally it involved *non-coding* DNA, and usually still does. But its swift establishment – in a little over twenty years – in a whole range of investigations provides a nice case study in how fast a new technology can be taken up, and how society has to be organized to make use of it. It is rapidly moving from a set of techniques used in a few specialized areas, whose demands have forced the pace of development, to a general-purpose identifier. The DNA ID card is still a little way off, but it is a real possibility. As one commentator on personal information put it a few years ago, "DNA is nature's digital signature".

How it works

The basic principles of DNA fingerprinting were figured out by the British scientist **Alec** (now Sir Alec) **Jeffreys** in 1985. He found a way to make analysis of what we now call the **genome yield information** of a new kind. In theory, it could record a pattern of DNA fragments which differ in every individual. In his original method, the differences say nothing about the working genes – although the term DNA profiling was used early on, it meant something different from the genetic profile screening tests might generate, though the distinction is nowadays harder to draw. The analogy with traditional fingerprints was apt. The DNA fingerprint revealed nothing about an individual except who he or she was. But, as with the older forensic technology, fixing identity could be very powerful.

However, there is a lot more to obtaining a DNA fingerprint than rubbing fingertips on an inkpad or dusting objects at the crime scene. And the process of interpretation is also more complex than traditional fingerprinting. It originally worked after the fashion of **Southern blotting**. Start with some biological material – blood, semen, hair or skin. The DNA, and a reference sample, could be chopped up with **restriction enzymes**, which recognize particular sequences. Alec Jeffreys' insight was that the lengths of the resulting fragments would be characteristic of each sample, due to variations in certain stretches of DNA where the same sequence was repeated lots of times. It was called the **RFLP** (pronounced "rifflip") method, short for **Restriction Fragment Length Polymorphisms**.

So Jeffreys was left with a collection of DNA pieces of different lengths. It was already established that, if you put mixed DNA into lanes in a rectangular slab of gel and passed an electric current through it, the charged fragments of DNA separated according to size: smaller ones moved up the lanes faster; and you could then blot the gel onto a nylon membrane, which picks up the DNA, and soak the membrane in a fluid containing specific, radioactively labeled DNA pieces, which bind to particular sites on the fixed DNA strands. These pieces – or "probes" – which matched the repeating sequences discovered by Jeffreys, were the key to the technique. The resulting pattern of radioactive bands was photographed on X-ray film. Comparing the dark bands across lanes of the gel, which developed like rungs of an irregular ladder on the film, showed whether the two samples were related, or even identical.

The whole procedure used to rely on the sharp eyes of the lab worker. And, in the early days, a **DNA "barcode"** needed quite a large sample, such as a whole drop of blood, and could take almost a week to prepare. It took considerable skill to avoid problems with degraded samples, uneven movement in different lanes of the electrified gel, and the like.

But it was immediately apparent how some people might benefit. The technique can be used to reveal plenty of other things besides identity. One obvious application was to aid mothers (usually) in bringing maintenance suits against an alleged father who denied paternity. The child's DNA is first compared with a maternal sample. This would reveal a match in half the bands in the barcode. The rest must come from the father, and a match with a putative father's sample would confirm his paternity. There have been one or two instances of the technique being used by "fathers" accused of incest to demonstrate that they were not the biological parent of the child concerned. It was also used in Britain in immigration cases wherein the authorities required proof of paternity. The Home Office soon accepted that DNA testing was more reliable than a blood sample.

Use in criminal cases was more various. Early successes included conviction of a number of rapists using DNA profiles from semen samples. One widely publicized case in the UK at the end of the 1980s – the first criminal conviction to be brought about based on DNA evidence – concerned the rape and murder of two teenage girls in Leicestershire. A 17-year-old boy, Richard Buckland, was charged with one of the murders, and confessed. But subsequent DNA testing – assisted by Alec Jeffreys and the Forensic Science Service – proved that he could not in fact be the culprit. (Proponents of DNA fingerprinting often point out that it is as important in establishing innocence as guilt.) In this case, the real killer, a local baker named Colin Pitchfork, was eventually found after 5000 local males gave DNA samples. He came close to evading testing by persuading a friend to impersonate him, but was arrested when the friend eventually told police of the deception.

DNA is not always the answer

Matching a DNA fingerprint can be a powerful piece of evidence, but a courtroom is not a laboratory. The DNA result can be contested, and becomes part of an argument which brings in other evidence. The 1995 trial of O. J. Simpson was a good early example of this.

Much of the discussion centred on drops of blood found alongside bloody shoe prints leading away from the bodies of Nicole Brown Simpson and Ronald Goldman, and the blood found on a gate at the back of the condominium where the murder took place. Blood from both places was a good match for Simpson on DNA analysis, which was repeated by three different labs. Simpson also had a cut finger when interviewed by police the day after the killings. The prosecution argued that the blood placed Simpson at the crime scene, where he had cut his finger with the murder weapon. It was not the only forensic evidence linking him with the crime, but its novelty made it some of the strongest.

The defence put this in doubt, suggesting that sloppy sample handling made the DNA results unreliable. They also suggested that the blood was planted at the crime scene after sampling from Simpson, an idea which other evidence of racism and corruption in the LA police department lent credence to. Others suggested that the prosecution's explanation of the ins and outs of the DNA techniques was poor. The jury found Simpson not guilty after four hours of deliberation.

Since these early cases, there have been thousands of uses of DNA evidence to establish guilt or innocence. The latter tends to be more definite. A matching DNA profile can only be interpreted as a *probability* – albeit a high one – that a particular person is linked to a sample. But no match means they are certainly not the person in question. As the technology has improved, it has allowed tests to be conducted on smaller samples, or on DNA that has been broken down into smaller pieces. Old evidence can now be revisited. In 2006, British prosecutors secured a conviction against a rapist whose crimes took place twenty years earlier, using a sperm sample recovered from one of his victims and preserved on a microscope slide. They tracked down the man using the **national database of DNA profiles** (see p.228).

DNA fingerprinting updated

The original **RFLP** method, which proved the potential of DNA profiling, has now been superseded by more sensitive techniques. The most widely used depends on **short tandem repeats** (STRs). As before, they use repeat stretches of DNA of three, four or five bases, which vary widely between people in the number of repeats at any particular place. Known STRs are tagged with specially made primers, and copied using the polymerase chain reaction now used in all DNA labs. This makes the procedure much more sensitive. It can use very small amounts of DNA, so stored material from old crimes can now often be analyzed to give new results. However, the polymerase chain reaction technique (PCR) will amplify any contaminants present just as effectively as it will amplify the original sample – working on older material makes more demands on lab workers. The DNA fragments made are again separated by electric charge – either on a gel or using a newer technique known as **capillary electrophoresis**.

The variations (polymorphisms) at any one STR region are very common, each appearing in between one person in twenty and one in five. But they vary independently, so testing many STRs can give much better discrimination. In a typical set-up, the preparation of DNA fragments is injected into a fine glass tube in which the pieces migrate through an inert polymer when a voltage is applied to the ends of the tube. They show up because the primers used to make them carry a fluorescent dye. The whole process is automated, and DNA profile analyzers read many fragments at once and compare them with a sample which contains all the standard variations at the SNP sites under test. In the UK, standard DNA profiling uses ten STR sites. Common practice in the US uses thirteen sites, which reduces the chance of mismatches still further.

The new-found sensitivity of the techniques has led to the overturning of old criminal verdicts, as well as the solving of old crimes. In the United States, for example, 120 men were released from jail between 1989 and 2002 – some from death row – after DNA analysis ruled them out as being contributors to biological samples linked to the crimes for which they had been convicted.

This technology is now widely available both to institutions and private buyers. Commercial companies commonly serve both, and will sell you a DNA test direct for a few hundred dollars or pounds – if you provide, as one suggests, bone, hair follicles, blood, semen, cheek cells, cigarettes, licked envelopes, toothbrushes or chewed gum. For private individuals, or their legal advisers, their main business is paternity testing. Typical figures from such operations are that between one in five and one in three of men involved in such tests turn out *not* to be the father of the child in question. Note that this figure is derived solely from those who were tested, so is a skewed sample: these men were tested for a reason. It is important to point this out because there is still little good information about how many children are not biologically related to their supposed fathers – journalists often recycle a figure of ten percent, but a British team's review of the scientific journals in 2005 suggested the real figure was more likely to be one in 25. Exact figures are hard to come by, but there may be 20,000 or more paternity tests annually in the UK, and 100,000 in the USA. As with other DNA-based tests, there is real concern that paternity tests are offered with no counselling or follow-up on what, for many, must be disturbing news.

Commercial DNA profiling for paternity testing is also used, less frequently, to prove a relationship to parents who may have lost contact with their children, and to secure an inheritance. And it is increasingly used for **genealogy**, both for people who want to trace ancestry to particular geographical regions or, on an even finer scale, to specific social groups. If you are of Scottish descent, for example, a recently compiled database of Clan Chieftans' DNA can give you a clue whether you are, as it were, the real McCoy. In other species, DNA profiling is also used to confirm pedigrees of show dogs and, with much more money at stake, racehorses. (There is also a wealth of scientific applications of the analysis of DNA markers in studies of evolution, which we discuss in chapter 5).

Establishing identity from human remains

Apart from crime, immigration, paternity and genealogy, DNA profiling has also played a significant part in establishing identity in cases where other evidence is lacking, or fails to convince. There have been a number of historical cases which have made headlines. DNA confirmed that a man found drowned off the coast of Brazil was the Nazi concentration-camp doctor **Josef Mengele**, who fled to South America at the end of the war. His remains had been buried in 1979, but the skeleton was exhumed and yielded enough DNA to compare with blood samples from Mengele's wife and son, still alive in Germany. It was the end of a forty-year hunt for the "doctor of Auschwitz".

Similarly, DNA has been used to confirm that remains unearthed from a bog in the Ural mountains in Russia in 1991 were those of the last Russian Tsar, Nicholas and much of his family, shot by the Bolsheviks in 1918 (though this analysis has been challenged by other scientists). On the other hand, tests on preserved tissue removed from the woman known as **Anna Anderson**, who had claimed all her life to be the Tsar's lost daughter Anastasia, failed to match the rest of the family. And if that seems an unlikely application of DNA profiling, tests on bone samples are in hand to try and establish whether the final resting place of **Christopher Columbus**' remains was in Spain or in the Dominican Republic.

These high-profile cases, while they may involve fragile or degraded samples, are the tip of an iceberg of much more complex identification. Wars and disasters nowadays generate long, painstaking and difficult DNA investigations which try and show exactly who the victims, and occasionally the survivors, were.

Again, the survivors tend to attract most attention. The scale of the crisis of the 2004 tsunami in Sri Lanka was hard to comprehend, but everyone could empathize with the nine mothers who claimed "**baby 81**", a tiny boy found washed up amid the wreckage. The hospital staff caring for the baby were understandably anxious not to give him to the wrong parents, so a judge ordered DNA testing to identify his actual parents, who were among those who had come forward to claim the infant, so one family could be reunited. However, the quick results from a straightforward case such as this are the exception. The aftermath of disasters and wars is usually a horrific mess, and even the most painstaking work, often over years or even decades, cannot answer everyone's questions. But DNA technology is still much more powerful than anything that has been available in the past, and great efforts are now expended to trace the fate of people caught up in some of the worst events of the past decades.

The first really large-scale operations were set up after the civil war in the former Yugoslavia, a horrific and bloody conflict. Many of the war's dead had been left in poorly concealed mass graves. The G7 Summit in 1996 agreed to set up the **International Commission on Missing Persons** (ICMP), based in Sarajevo. By mid-2006, after ten years' work, it had managed to recover DNA samples from 14,000 victims, and taken a further 80,000 samples from families hoping to learn the fate of someone missing. This databank enabled them to identify 10,000 of the people killed, and match them with their families. Nonetheless, it left 27,000 more missing people unaccounted for. The ICMP has since become involved in other war and disaster zones, sometimes helping with direct analysis, sometimes with advice on technique. You can study its work and its results at www.ic-mp.org/home.php. The details often make harrowing reading.

Perhaps even more challenging than mass graves of intermingled bones in Serbia was the gruesome task of identifying remains from those who died in the destruction of the **World Trade Center**'s twin towers in 2001.

Although there is a fairly accurate list of the victims of 9/11, many of the bodies of those trapped in the towers had been pulverized. As the clearance advanced, more and more human body parts were found, many of them mere fragments. Investigators found that DNA obtained from them had often been damaged by the intense heat of the fire, but mitochondrial DNA matches, using sequences that exist in many more copies, have helped in some cases. Five years after the atrocity, New York's medical examiner's office had recorded over 20,000 body parts, and carried out over 100,000 DNA tests. But these had still only led to positive identification of 1600 victims. Another 1150 families were still waiting for a result.

The 2004 tsunami in Asia was on a different scale again. It resulted in the deaths of over 150,000 people, and generated the largest operation for storing and matching DNA after a disaster so far. It was often the case that people had died alongside their relatives, and personal items such as hairbrushes, which can yield DNA samples, were mostly lost. The scientists working on the victim's remains had to develop computer programs to match distant relatives. Much of the DNA identification effort, however, focused on victims from the West, whose countries had the resources to pursue identification. Even then, DNA was mainly used as a supplement to dental records and fingerprints. The ICMP was involved in both these operations, and it has also advised the government of Iraq and provided help to the authorities in Louisiana after Hurricane Katrina.

DNA detective work is now being applied to the still continuing effort to identify victims, and to reunite Jewish families dispersed during the Holocaust. The **DNA Shoah** project, announced in 2006, uses tools developed after 9/11 and the 2004 tsunami to work on a new database of the DNA of Holocaust survivors. Some may have profiles which match those of Holocaust remains recovered from central Europe – for example, from land development projects in Poland and Germany. Others are probably related to the 10,000 Holocaust orphans who came out of Europe after the war. There are 300,000 Holocaust survivors, most heading into their 70s, and the DNA sampling is only just beginning.

Personal DNA databases

It is all very well having DNA from a crime scene. But you need samples to compare it with. These can be worked up for the case, as in some of the famous examples above. But criminal investigation might be easier if there was a large collection of profiles already in place.

Now there is, at least in the UK. The country where DNA fingerprinting was invented was also the first to establish a **National DNA Database** (NDNAD). Not to be confused with medical research collections, such as the more recent **Biobank** (see p.83), it was established in 1995 to hold samples collected from people in custody. The circumstances in which samples can legally be taken have gradually been widened since then. Under the latest UK legislation **non-intimate samples** – usually a cheek swab – can be taken without consent from anyone who finds themselves in a police station during investigation of a "recordable offence" – basically, anything more serious than a traffic violation. The sample can be taken even if it is not relevant to the crime under investigation.

The database's operators run a "speculative search" on every new profile, whether from a known individual or a crime scene, to see if there are any matches with profiles already collected. (There is a separate collection of profiles of the police, in case they get careless with their own DNA at a crime scene, but this is not speculatively searched.) In England and Wales (but not in Scotland), the profile obtained can be kept on the database even if the person is never charged, or is charged and acquitted. The DNA samples are also retained, but by one of the six laboratories doing the profiling, not centrally. The whole operation is overseen by the **Forensic Science Service**, a government-owned company, a separate "custodian" which is based in the Home Office, and a Strategic Board.

This has cost a great deal – more than £200 million since 2000 – and now covers a lot of people. In 2006, the NDNAD had DNA profiles of over three million people – up from 2.4 million in mid-2004, and ten times the proportion of the US population who had then been profiled by law enforcement agencies. And it looks as if those three million people cover a large proportion of offenders. DNA profiles are produced from less than one in a hundred recorded crimes in the UK – there has to be an actual crime scene, some DNA has to have been left, and the police have to recover it. But if all three things happen, there is now a 45 percent chance the crime profile will match one already in the database. If it does, then any use of DNA evidence in court relies on a new sample taken from the accused, not on the old profile.

There are concerns, however, about the operation of the database. The Home Office says people who are sampled but who never go on to commit an offence have nothing to fear from the retention of their profiles – the same argument that is routinely used to reassure people bothered about the increasing use of CCTV cameras. Its opponents see the database as tagging people who will then be treated as permanently under suspicion. As the **Bishop of Worcester** put it in the House of Lords in 2003 in a debate on the Criminal Justice Bill, it creates three categories: "the guilty who have been convicted of offences, the not guilty, and the probably dodgy". A serious worry here is that the category of the "probably dodgy" is likely to include a disproportionate number of people from ethnic minorities.

There are some tricky additional issues, too. If there is no direct match, for example, the DNA database can be searched for near-relatives of the person whose DNA has been recovered. Some of these people may not even know they are related until they become involved in an investigation entirely unconnected to the one they were originally called in for.

There is also the possibility of incorrect matches. What are the chances? The official estimate of the likelihood of two unrelated profiles matching is less than one in a billion. However, DNA fingerprinting pioneer **Sir Alec Jeffreys** thinks that the current profiles – which record ten STR markers and a gender marker – should be improved to cover fifteen markers. The police are also interested in taking the analysis beyond straight DNA profiling and using our more detailed knowledge of the human genome to derive likely physical features – such as hair, skin and eye colour, stature and even facial appearance. This would be a "genetic photofit" rather than a genetic fingerprint. But the technology is currently a long way from any such development.

Other possibilities include new equipment for crime-scene DNA analysis. First a lab in a van, perhaps, and then eventually a lab on a chip. But could this be done with the kind of quality control now in place in fixed labs? There are also new forms of sample collection. For instance, bus drivers, traffic wardens and train inspectors in various parts of the UK have been issued with simple "**spit kits**" – a pair of latex gloves, a collection bag and sterile swabs. At least one offender has been convicted in Manchester for spitting on a traffic warden who kept a sample which was sent for DNA analysis. There may also be moves toward remote access to the database. Some have suggested these will need to be watched closely to ensure validity of evidence, and confidentiality and security of any transmitted data. And as crime – and terrorism – are not respecters of national boundaries, there are plans to share data between databases in different countries, which may not have the same procedures in place for quality control or security.

So far the UK has led the way, from the first stage of DNA technology in criminal investigation involving visual comparison of profiles obtained during work on a particular case, to the current possibility of fishing for suspects, or even their relatives, in large, computer-searchable databases. At the moment it looks likely that more countries will adopt the British model. The US is expanding DNA profiling as part of a $1 billion initiative announced by President **George Bush** in 2003 to build on the Combined DNA Index System CODIS, which dates from the late 1980s. It is paying for updating lab equipment, training and research and development to help eliminate a national backlog of unanalyzed DNA samples from crimes. The money is being parceled out to individual states. You can read about recent developments at www.dna.gov

A bill approved by the **US Senate's Judiciary Committee** in 2005 provided for the legality of collection of DNA samples from suspects by the Federal authorities, and for their retention whatever the outcome of the case. However, it would still allow those arrested to petition to have their details removed from the database when the case was closed.

No other European country yet comes anywhere near the UK in collecting DNA profiles of its citizens – criminal or otherwise. Although most now have legislation in place to allow collection of samples, they generally number in the low thousands at most, and their use may be restricted to serious crime. However, a recent academic study suggested that there is a clear trend for any country that has a database to gradually expand the categories of people from whom samples can be taken and retained. No one, as yet, has ever built a database and then reduced its use.

There is another set of justifications entering these discussions, too. Governments are looking for secure **biometric data** to identify who has citizenship rights, and to track movements across national borders. These proposals now tend to involve fingerprints, or the more recent technology of iris scans. But DNA is seen as the gold standard. It seems quite likely that by mid-century there will be national, or even pan-European **identity recording** that will include a DNA profile for every citizen. If you want to track these kinds of development, check out www.dnaresource.com, an excellent website run by a law firm specializing in forensic uses of DNA.

PART 5
GENETIC
FUTURES

Stem cells

The promise of renewal

Like genes, stem cells are rarely out of the news these days. Because of all the hype and controversy, it's sometimes difficult to work out exactly where stem-cell research is at, and where it's going. Most scientists would agree that cell-based therapies offer one of the best hopes for treating incurable conditions such as spinal injury, Parkinson's disease and diabetes. This approach, a logical extension of existing transplant operations, is called regenerative medicine.

But cell therapies, like genetic therapies, are still in their infancy. Much more work needs to be done before it becomes clear which type of cells and techniques is going to bear fruit, or if there will be any unforeseen drawbacks. In this chapter, we'll look at why stem cells are so special, and why some aspects of stem-cell research have proved so controversial.

Stem cells – the basics

A stem cell is one that can both **self-renew** and give rise to more specialized cells. They are found everywhere, at all stages of development and probably in all tissues. Stem cells give rise to all the different cell types in the developing human embryo, which starts off as an amorphous blob, and eventually forms a complex organism made up of over 200 different types of tissue. Throughout life, stem cells are the body's "repair kit", responsible for replenishing cells lost to wear and tear. So **skin stem cells** make more skin, **blood stem cells** make red and white blood cells, **muscle stem cells** make more muscle, and so on, for as long as the person is alive. Research on stem cells that have been isolated from embryos is shedding light on embryo development, while work on other types of stem cell is helping scientists understand how cells replace damaged or worn-out tissue in adults. A lot of effort is going into finding out how these processes can be replicated in cells growing in a laboratory dish, with the eventual aim of growing custom-made tissues – and perhaps even whole organs – for transplantation. Such cells will also be useful for testing the effects of new drugs, and in studying development in embryos affected by disease.

Stem cells at different stages of development have different degrees of "potency", that is, their potential to give rise to other types of cell. Stem cells isolated from very early embryos, called **embryonic stem** (ES) cells, are "**pluripotent**" – in the right environment, they can grow into any of the body's cell types. So-called **adult stem cells**, on the other hand, are "**multipotent**", as they give rise to a more limited range of cell types. For example, blood stem cells (haematopoietic stem cells, which are found in the bone marrow), make white and red blood cells, but not brain or muscle cells. A cell that has specialized is described as differentiated – a process that scientists used to think was irreversible. But then came **Dolly** – a whole new sheep, grown using the genetic material of an udder cell. Turning back the clock, or "reprogramming" specialized cells in the laboratory, so that they regain the ability to produce many different cells, is a very promising area of stem cell research. Several recent studies also suggest that specialized cells may be capable of switching jobs in later life, if provided with the right nurturing environment – currently another very active line of scientific enquiry.

Embryonic stem cells

Scientists discovered how to extract stem cells from mouse embryos and grow them successfully in the laboratory over twenty years ago. But it wasn't until 1998 that a US team led by **James Thomson**, of the University of Wisconsin-Madison, reported the first successful isolation of *human* ES cells. The therapeutic promise of this achievement was not lost on those working in the field, who quickly realized that if these cells could be multiplied in the lab, and persuaded to grow into different tissues, they could potentially provide new treatments for a whole host of conditions – insulin-producing cells for diabetes, brain cells for Parkinson's disease or nerve cells for repairing spinal injuries, to name just a few.

Such cell transplants would need to be carefully tissue-matched (as with blood transfusions), to avoid rejection by the patient's immune system. But again, there was a potential solution to this problem, made possible by the cloning technology used to create Dolly the sheep: "**somatic cell nuclear transfer**", or SCNT (see chapter 10). So-called "**therapeutic cloning**" could potentially allow the isolation of stem cells from cloned embryos created using genetic material from the patient's own body cells. This would involve removing the nucleus (which contains the genetic material, **DNA**) from, say, a patient's skin cell, and placing it into an unfertilised human egg emptied of its own genetic material. In theory, such genetically matched cells could be used to replace any

Developmental biologist James Thomson in his Primate Research Center lab.

dead, diseased or missing tissue. In practice, many hurdles remain, and at the time of writing no one has yet managed to grow stem cells isolated from cloned human embryos.

Challenges of ES cell research

Isolating human ES cells requires a **human embryo**, which is destroyed in the process. This fact has led to such research being banned or limited in many places, and is the focus of an ongoing, highly charged debate (of which more later). Most of the ES cells isolated so far have been obtained from "spare" embryos donated by couples who have undergone **in vitro fertilisation** (IVF) treatment, although some countries, including the UK, also allow scientists to work on early embryos specifically created for research. To get at the ES cells, an IVF embryo is first grown in a dish for three to five days, by which time it has become a hollow ball of cells called a **blastocyst**. Inside this ball is a clump of around 30 cells, called the inner cell mass, which – if the embryo were returned to the womb and allowed to develop – would eventually give rise to all the different organs and tissues of the body. The inner cell mass is carefully removed from the embryo, and placed into a plastic dish containing a cocktail of nutrients and "growth factor" proteins.

With luck, the isolated cells will start to divide and multiply in the dish, eventually covering its surface, at which point they are gently removed and split into several more dishes – a process known as **"passaging"**. By repeating this many times, after several months the original 30 cells will have produced millions more. The ideal result is that all these progeny will be unspecialized ES cells, still capable of replenishing themselves – if so, then the ES cells are now described as an **ES cell line**. As well as checking the physical appearance of the cells under a light microscope, there are certain other checks that scientists can do to verify that ES cells are still completely unspecialised – for example, measuring the activity of certain genes known to be switched on during very early embryo development. Given that many human embryos contain genetic errors that prevent them from developing into a foetus, it also makes sense to check the genetic integrity of any new ES cell line, as far as possible.

This is the simple version of the procedure. In practice, cell culture is an area of laboratory work which can still be quite hit-and-miss – in some ways it is more like cookery than a set of precisely controlled experiments. And it is tricky to maintain pure cell lines with defined characteristics as cells often grow and change in the culture. This needs to be borne in mind in the face of the excitement surrounding stem-cell research, which has attracted much attention as the next big thing in biotechnology.

Apart from obtaining high quality cell lines, much remains to be done before ES cells can be used therapeutically. As we explained in chapter two, the fate of a particular cell in a developing embryo – whether it becomes a toe, liver or ear cell – depends on its location, the instructions it receives from other cells and those it receives from its own genes. Recreating this complex set of signals in the laboratory is no mean feat, but some progress has already been made towards this goal. To some extent, this process happens unaided – if ES cells are left to clump together then they will naturally form "**embryoid bodies**", which may be made up of muscle, nerve or any other specialised cell type. By fiddling about with the basic growth cocktail (properly called **cell culture medium**), scientists have come up with tailored "recipes" that encourage the growth of particular cell types. But research is needed before cell specialization can be strictly controlled, producing pure populations of the tissue required.

A further issue with most human ES cell lines is that they have been grown on top of a "**feeder**" layer of mouse cells, which provides essential support and nutrients. These cells will never be suitable for use in human treatments, since the risk of infection by animal viruses or rejection by the patient's immune system would be too great. Much effort is now going into the creation of new "clinical grade" human ES cell lines, grown under strictly controlled conditions and without using any animal products.

A stem-cell scandal – and Snuppy

As mentioned earlier, one of the aims of ES cell research is to investigate the possibility of creating cell lines tailored to an individual's genetic make-up, using cloning technology. Such work is technically very difficult, with a low success rate, and is also dependent on a plentiful supply of human eggs. Obtaining eggs from women's ovaries is an invasive, potentially risky procedure, and as such there is a shortage of donated eggs for fertility treatment – let alone for medical research. But, in February 2004, it seemed that a team based at **Seoul National University** in South Korea had managed to overcome these seemingly insurmountable difficulties – they obtained one cloned ES cell line from 30 cloned human embryos, after more than 200 tries. Previously, US and UK scientists had managed to clone human embryos, but they hadn't got them to survive long enough to isolate stem cells. **Woo Suk Hwang** and his colleagues seemed even further ahead of everyone else when, in May 2005, they reported making 11 cell lines from 31 cloned embryos, using just 185 eggs. What's more, the cell lines were supposedly created using cells from patients with spinal injuries, a genetic immune disorder and Type 1 diabetes. This astonishing achievement, an apparent ten-fold increase on the team's previous cloning attempts, seemed too good to be true – and, sadly, it was.

Hwang had put his success down to the skill of his research team, and the use of fresh eggs, provided by fertile donors. But it was questions over the ethically dubious sourcing of these eggs, and suspicions over the actual numbers used, that first lead to Hwang's achievements being scrutinized further. Towards the end of 2005, it was revealed that some of the donors had been paid, while some eggs had come from junior team members. These allegations were swiftly followed by the astounding news that the team had fabricated the existence of its cloned human ES cell lines. The scientific papers describing the advances were hurriedly retracted, while Hwang was dismissed from his university post and charged with fraud and embezzlement of government funds (the trial is ongoing as we write).

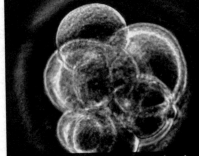

A cloned human embryo, created at the Centre for Life in Newcastle upon Tyne, England, three days after the nuclear transfer took place.

The one consolation for the disgraced researcher was that, despite his spectacular fall from grace, extensive tests on **Snuppy** – the world's first cloned dog, also created by the SNU team – showed that he was a genuine clone. Snuppy was created using the genetic material of an ear cell taken from a three-year-old male Afghan hound, and was the only one of over a thousand cloned embryos transferred to surrogate dog mothers that survived. The scientists followed this up with the cloning of three female Afghan hounds, also created using DNA from an ear cell. The births of **Bona** (Latin for blessings), **Peace** and **Hope** were announced in January 2007. Given his ability to pull off a procedure that continues to confound other cloning researchers, there could always be a future for Hwang in the pet-cloning business.

Groups opposed to all ES cell research were quick to point to the Korean ES fraud as evidence that such work should be stopped, even before it had really started. But this was to discount all the genuine work being carried out by every other stem-cell scientist around the world. The promise of new cell therapies for a host of different diseases remains – although whether they will come from ES or other types of stem cell is still unknown. It may be that public banks of ES cell lines – operated on a similar basis to blood banks – might offer a more realistic hope for clinical applications than individual, genetically matched cells. It has been estimated that ES cell lines from just ten carefully selected individuals could potentially provide matching tissue for around eighty percent of the UK population.

It could even be that ES cells will never be used therapeutically, but instead will be mined for rich seams of information on poorly understood conditions such as **motor neurone disease** (MND). This is the aim of Dolly's creator, the scientist **Ian Wilmut**, who holds one of the few UK licenses to create cloned human embryos. Wilmut wants to produce ES

Snuppy (right), and the male Afghan hound that provided the adult skin cell used to clone him

cells from cloned embryos created using skin cells from MND patients. These cell lines could then be studied in the laboratory for signs of the debilitating neurodegenerative disease. But for Wilmut and other stem cell researchers, there is still the pressing issue of where the human eggs needed for the work will come from. The UK's **Human Fertilisation and Embryology Authority (HFEA)** is currently considering whether women should be able to donate eggs for medical research at any time. At the moment, British women can donate eggs for fertility treatments, but are not able to donate them for research unless they are already undergoing a fertility treatment or sterilization procedure. Such eggs are few and far between. Researchers can also use eggs that failed to fertilize during the IVF process, but these are likely to be of poorer quality, and have genetic errors that prevent normal development.

Some stem cell researchers feel that asking women to donate eggs for cloning research is both unfeasible and unreasonable until the efficiency of the process can be improved – at the moment, it is likely that hundreds of eggs would be needed to create just a few cloned ES cell lines. So another possibility is to use eggs from another mammal, for example **rabbits or cows**, to allow scientists to hone their technique, and also to create cell lines from patients for research – though, obviously, not for treatment. This would involve placing the genetic material from a human body cell into a rabbit or cow egg that had its own nucleus removed. Two teams of UK stem-cell researchers recently asked for permission to take this approach, in the hope of overcoming the egg shortage problem. However, perhaps worried by newspaper headlines about "Frankenbunnies", a recent Government White Paper proposed banning the creation of embryos that contain a mixture of human and animal material. The HFEA is running a public consultation on the matter, so the research is currently on hold

Adult stem cells

One way to get around some of the problems associated with ES cells is to turn instead to "adult" stem cells. In fact, an adult stem cell refers to any self-renewing cell that can also produce a limited range of more specialized cells, whether it is present in the foetus, child or adult. For this reason, some scientists now refer to them as "**somatic stem cells**", rather than adult stem cells. One sort of adult stem cell – those found in bone marrow – has been used in research and medical treatments for decades. Bone marrow actually contains two types of stem cell: **haematopoietic,** or blood stem cells, which make all the different types of blood cell, and **stromal cells**, which form bone, cartilage, fat and connective tissue. Bone marrow transplants have been used for thirty years to treat conditions that affect the blood, such as leukaemia, inherited immune disorders or severe anaemia. As usual, there has to be a match between donor and recipient for certain key cell-surface proteins, as appeals for potential donors for waiting patients frequently testify.

Although adult stem cells are thought to be present in every tissue, their numbers are small, so getting hold of them is tricky. And even once they are sitting in a dish, adult stem cells are notoriously difficult to grow in the lab. As with ES cells, scientists are trying to find ways in which they can coax adult stem cells into producing specific tissues in the lab. One startling discovery made in recent years is that some adult stem cells may be able to produce specialized cells of other tissues. There have been reports, for example, of blood stem cells growing into brain, muscle and liver cells. It's not yet clear whether this phenomenon – called "**transdifferentiation**", or plasticity – is something adult stem cells do naturally in the body, or if it only happens under certain laboratory conditions. If scientists can identify the mechanisms and chemical signals behind this cellular alchemy, then the possibilities for repairing a patient's tissues using their own tissues may be limitless.

Whether stem cells come from embryos or adult tissues, before they can be used in new therapies, scientists need to find out how to grow them in large numbers in the lab and how to reliably direct their growth into the specific tissue required. They then need to find ways of making sure they survive in a patient after transplantation, and do the job they are supposed to do. Most importantly, they need to be sure that stem cell therapies will not have any unwanted side effects, such as cancer or infection. In short, many years' work lie ahead to get stem cells from the bench to the bedside. Some promising early successes have been claimed from giving patients with heart failure injections of their own stem cells into heart muscle, but many further trials are needed before this becomes a proven therapy. Meanwhile, UK stem-cell experts recently urged people to stay clear of unproven stem cell "cures" already offered by clinics worldwide for conditions such as multiple sclerosis. They are worried about patients getting ripped-off, as well as anxious about the reputation of the field.

The cord blood cell controversy

As we've said, bone-marrow transplants are already used to treat many different diseases. In some cases, rather than using blood stem cells obtained from bone marrow, doctors now use those from the umbilical cords of newborn babies. However, cord blood does not yield as many stem cells as bone marrow, so is used more often for operations on children than adults. Nevertheless, the real and perceived potential benefits of stem cells have lead to a number of private firms offering **cord-blood banking** services to new parents. This "life-giving opportunity that happens only at the time of a birth", as one of these firms puts it, will set you back a few hundred pounds. But whether or not this is money well spent is still hotly debated. Critics say that such companies play on parents' fears, as they offer future speculative treatments for many conditions not currently treatable with cord blood, while supporters say that, until cord blood collection is routinely offered to all couples, the private sector is providing a valuable service.

The UK's **Royal College of Obstetricians and Gynaecologists** (RCOG) recently weighed in on the debate, saying there is little evidence to recommend the practice on an individual basis, and calling for increased funding into the NHS cord blood bank. Currently, this public bank collects between 1700 to 2000 samples each year, which are available to anyone who needs a transplant (and for whom a matching sample is available). Cord-blood storage is also specifically offered to families affected by genetic disorders that may be treatable with a transplant. But with recent reports suggesting that cord-blood stem cells may be able to produce other types of specialised cells, as well as blood cells, the cord-blood banking companies look set to continue doing brisk business. Sir Richard Branson may have come up with a solution that keeps everyone happy, however: his Virgin Health private cord blood bank, launched in February 2007, will split each collection in two. Half the cord blood will be kept for the child, while the other half will be deposited in a public bank for anyone to access.

The US stem-cell debate

From a scientific point of view, to fully understand the processes involved in embryo development and cell specialisation will require a thorough investigation of all types of stem cell, from all developmental stages and all body tissues. In the UK, stem-cell research benefited from regulations in place that already permitted the use of human embryos for research into infertility and contraception, provided they are destroyed before they reach fourteen days old (see chapter 11). If this regime is accepted, it is hard to argue against some of those embryos being used for work on stem cells; it is sufficient to ensure that embryonic cells are treated as special, and that work is carefully controlled. As before, the research has to be licensed by the **Human Fertilization and Embryology Authority** (HFEA).

However, in some countries – notably the US – embryo stem-cell research has become such a political hot potato that much effort and money is currently going into the search for *alternatives* to human ES cells. Such work aims to find ways of creating "embryo-like" stem cells that have the same therapeutic and research potential, but which do not upset the sensibilities of those opposed to destructive embryo research. Although led by anti-abortionist activists, the anti-embryo research lobby does not encompass all those who oppose abortion – several pro-life US politicians have publicly supported human ES research. They argue that early embryos, unlike foetuses, are not sentient, and that it would be more unethical to block research that could potentially prevent human suffering.

US researchers working in the private sector can carry out as much human-embryo research as they like, but those funded by federal money can only work with human ES cell lines already in existence on August 9 2001. At that time, when President **George Bush** announced his policy, it was estimated that 78 ES cell lines were available to US researchers, although later evidence showed that the actual number was closer to 20. Furthermore, none of these older cell lines will ever be suitable for therapeutic use, as all were originally grown using mouse "feeder" cells. Unsurprisingly, there was strong opposition to this restriction by many scientists and patient groups, as well as celebrities who stood to benefit from the research. They included the late actor **Christopher Reeve**, who was paralysed following a riding accident; **Nancy Reagan**, whose husband Ronald died of Alzheimer's disease in 2004; and star of the *Back To The Future* films, **Michael J. Fox**, who has Parkinson's disease.

The efforts of the pro-ES cell-research lobby eventually lead to a bill that would have lifted restrictions on federally-funded research, allowing scientists to derive new cell-lines from spare IVF embryos. The **Stem Cell Research Enhancement Act** was passed by the House of Representatives in May 2005, by 238 votes to 194, and the Senate finally approved the measure in July 2006, by 63 votes to 37. But these results counted for little, as Bush promptly exercised his Presidential right to veto any bill passed by Congress, for the first time since he came to power in 2001. The bill was reintroduced into the House as the Stem Cell Research Enhancement Act 2007, and although this time it was passed by 253 votes to 174, this is still short of the two-thirds majority needed to override another promised veto. In the meantime, a number of US states – notably California – now have legislation allowing state funds to be used in ES cell research, including the creation of new cell lines. In 2004, Californian voters passed

Proposition 31, which provides for the state to raise $3 billion from bonds which will finance almost $300 million worth of stem research annually. The measure was passed with support from scientists, patient groups, and biotechnology companies, all of whom feared the US might fall behind in a crucial area of research.

One of the first to promote ES cell alternatives to bypass the ethical controversy was US bioethicist **William Hurlbut**, who suggested a method he called "altered nuclear transfer". This and other approaches were put forward in a 2005 report published by the President's Council on Bioethics: "Alternative Sources of Pluripotent Stem Cells", available at www.bioethics. gov/topics/stemcells_index.html

Hurlbut's method is based on that used to create cloned embryos: scientists would remove the nucleus from a donor cell and place it into a human egg cell that has had its own genetic material removed. But by genetically manipulating the donor's DNA before it is inserted into the egg, the resulting cells could be prevented from organizing themselves into a viable human embryo. He based his idea on mouse embryos that lack a gene called **Cdx2**, which fail to develop a **trophectoderm** – the tissue from which the placenta normally develops – but which still produce embryonic stem cells. Hurlbut says that if human embryos were created that have the same genetic error, they would be an ethical source of stem cells as they do not have the potential to develop into a human being. Not everyone is convinced by his arguments – embryo stem cell researchers say that creating such "altered" embryos would be a costly and time-consuming diversion, since no one knows if knocking out the human version of the Cdx gene would have the same effect. And many of those opposed to human embryo research are unlikely to welcome the deliberate creation of embryos that are doomed to die.

Another approach, put forward by US firm **Advanced Cell Technology** (ACT), is to try to grow ES cell lines using single cells taken from very early IVF embryos. As with the technique of **pre-implantation genetic diagnosis** (PGD) – which involves removing a single cell from a 3–5 day old embryo for genetic testing – the method potentially leaves the embryos intact and able to develop if returned to the womb. But again, this method has its drawbacks: who would actually choose to subject their IVF embryos to this procedure, unless there was a pressing medical reason? Furthermore, the scientists faced criticism from both sides of the divide, when it emerged that all sixteen of the human embryos used in their "proof of principle" study had actually been destroyed.

Yet another alternative comes from scientists in Italy, which has some of the world's strictest laws on human embryo research. A team from the University of Milan has managed to create human ES cell lines from **"parthenotes"** – eggs that have been rigged to start dividing without fertilisation by a sperm. Human eggs contain two sets of chromosomes until fertilisation, when the entering sperm triggers the second set to be expelled. In order to create human parthenotes, researchers must block this expulsion and stimulate the egg to divide. As some of the genes required to drive development are provided by the sperm, such "embryos" usually die within a few days. This means that parthenotes are not capable of forming independent human life and so, the researchers hope, would sidestep ethical objections to ES cell research. Finally, the possibility of using "arrested" IVF embryos – those that have stopped developing in the lab, but are still "alive" – as a source of ES cells has been raised. Again, whilst potentially useful for basic research, such embryos are likely to contain serious genetic errors that stopped them growing in the first place.

It is unlikely that any of the above methods will please everyone – scientists would rather work on ES cells that come from genetically normal, healthy embryos, whilst many pro-life activists oppose any technology that involves manipulating eggs or embryos in the lab – including routine IVF. Most of those opposed to ES cell research are promoting adult or cord blood cells as the only genuine ethical alternative (although, as we outlined earlier, there are practical limitations with both of these). But, for scientists living in countries such as Germany and Italy, where all human embryo research is forbidden, "alternative ES cells" may provide a way to advance understanding of stem cells without breaking the law.

It is hard to predict how, and how fast, the field will develop. Understanding of cell biology is increasing all the time, which should improve the prospects for getting ES cells to do what doctors want, as well as making it easier to use stem cells from other sources. At the same time, countries eager to develop in biotechnology, such as China, have invested heavily in stem-cell research in the hope that restrictions elsewhere may enable them to leapfrog their competitors. The **World Stem-Cell Map**, compiled by an activist, displays policies worldwide, and its author reckons that more than half the world's population live in countries with permissive or flexible policy on human embryonic stem-cell research. Apart from the UK and China, they include Australia, Canada, India, Iran, Russia, and South Korea. Stem cells, it appears, are here to stay.

Future health care

Personalized medicine?

While it was clearly an exciting challenge for the scientists, much of the enthusiasm for reading the human genome came from forecasts of the significant impact it would have on health care. As we saw in chapter 7, genes play a valuable part in contemporary medicine, but it is fair to say that much of the science's impact still lies in the future. And it is hard to weigh up, as the predictions of the gene pundits vary widely. There is certainly something in the often-heard suggestion that detailed knowledge of each individual's genetic make-up will allow a shift toward "personalized medicine", for treatment and perhaps prevention of conditions under genetic influence. But how much, and how soon?

Some are confident it will happen fast. *The Economist* quoted the American biologist and genome proselytizer **Leroy Hood** in 2005: "Ten years from now, you will not have to spend hours in a doctor's office to complete a comprehensive health check-up. Instead, with just a single pin-prick, a nanotechnology device will quickly measure and analyze 1000 proteins in a droplet of your blood. Based on this "molecular fingerprint", your doctor will prescribe drug regimens tailored to your personal state of health that will not only be able to reverse many diseases, but will also prevent their manifestation in the first place."

Let's investigate this claim. Hood is a visionary, and his vision is based on the development of the biology that makes genomics possible (rather than genomics itself). This "**systems biology**", he believes, will eventually allow us to untangle the interactions between all the molecules in a cell or tissue. That in turn will lead to understanding how they operate in complex networks, and how these networks are altered in disease. Factor in the development of technologies that allow the state of the networks to be read almost instantaneously, and the new predictive, preventative, personalized medicine he anticipates finally becomes possible. In his scientific papers, he puts it a bit more cautiously than when he is speaking to *The Economist*: "If gene expression in diseased tissues ... reveal[s] patterns characteristic of pathological, genetic or environmental changes that are, in turn, reflected in the patterns of secreted proteins in the blood, then perhaps blood could serve as a diagnostic window for disease analysis".

At his **Institute for Systems Biology** in Seattle, Hood is working to realise this vision, and believes that both the tools and the analysis will come together in a decade or two. Realistically, it will take longer than that. Here, we look at some of the nearer-term possibilities, which still focus on using the information retrieved from our DNA. Then we look at the more far-reaching ideas that would only come to fruition if we decided to change our genetic make-up – and so, in a sense, change ourselves.

Personal prescriptions

Pharmacogenetics

When the doctor gives you a pill or a jab, the process would ideally go something like this. A chemical is released which targets the appropriate cells in whichever one of your tissues is causing you a problem – and *only* those cells. It blocks whatever is doing the damage, or puts it right. There is enough of it, and it lasts long enough to finish the job. Then it goes away without doing anything else.

This never happens, of course. Drugs, by definition, are biologically active. They may do things the doctor did not intend. And once they are inside you, your cells may get to work on them and complicate things. The medicine you are given may end up converted into a more active form. It may encounter an enzyme or two that turn it into something harmful. It may need to be transported to a particular part of the cell to do its job. Or it may just break down and be excreted before it has done anything much at all.

All these things, as well as the actual target for the drug when it is in the right form, are likely to be influenced by variations in the relevant genes. So knowing more about those variations ought to help doctors choose the right drug, at the right dose. The possibility first emerged around fifty years ago. Some African-American soldiers who took an anti-malarial drug called **primaquine** became very anaemic. They were found to have low levels of a crucial metabolic enzyme, **glucose-6-phosphate dehydrogenase**. The same metabolic quirk can lead to damage to red blood cells (causing anaemia) from other drugs, including aspirin.

Some other widely used drugs, including the anti-tuberculosis medication **isonaizid** and the anaesthetic **succinylcholine** were also found to have variable effects due to differences in individual biochemistry. The idea of a more general study of these, **pharmacogenetics**, was first put forward at the end of the 1950s. It simply means the study of genetic variation that affects response to medicines. Those who want to impress nowadays call it pharmacogenomics, which means pretty much the same thing, only with more genes. The term was used a great deal in the years which saw the political, commercial and medical excitement around the last stages of the **Human Genome Project**. But it was used more as a shorthand for big expectations than as a way of referring to anything concrete in the way of results. Sussex University sociologist Adam Hedgecoe reports that between 1997 and 2000 there were more review articles – professional overviews of scientific literature – using the term **pharmacogenetics** than papers presenting original research.

It is true, though, that we now know a lot more about some of these genetic effects, and often understand them in great detail. There are hundreds of variants of the gene coding for glucose-6-phosphate **dehydrogenase**, for example. As the genome project developed, the catalogue of genes known to affect drugs grew, which boosted hopes for the possibility of personalized prescriptions based on individual gene profiles. Some of these would only be needed to avoid making things worse by using the wrong drug. More positively, there are likely to be new classifications of disease based on varying individual genetic characteristics among patients with what are now thought of as single medical conditions.

A simple example of how this might work is the recent use of the anti-cancer agent **Herceptin** (trastuzumab). This is a monoclonal antibody which binds to a protein called **human epidermal growth factor receptor 2 protein** (also known by its significantly less cumbersome acronym **HER2**). Around a quarter of women with breast cancer have tumour cells that over-produce HER2, and the antibody can either kill these cells, or aid their destruction by the immune system, or other drugs given to the patient at the same time. But although it is often cited as an example of the new genetic approaches, the prescription is only pharmacogenetic in the sense that patients are tested to see if they are over-producers. The alteration is in the cancer cells, and has nothing to do with inherited change.

The drug, and the test, were first licensed in the USA in 1998. For those who received Herceptin in early-stage breast cancer, trials so far have shown that the chances of the cancer returning are roughly halved, though there are also side-effects including occasional heart problems. In the UK, there has also been fierce controversy about the cost of the drug, with some patients suing health authorities who declined to fund the cost of treatment.

The fully developed idea of personalized medicine would go well beyond testing a patient already diagnosed to see whether they ought to have a particular drug. It would depend on a full genetic profile, which could feed into advice about prevention, prognosis and, if necessary, prescription for a range of conditions.

This futuristic scenario is predicated on the idea of a "**gene chip**", a microarray that scans many gene regions at once and records the whole lot for assessment. These are coming into wider use, although tests of their reliability are still under way. They are the offspring of a marriage between DNA science and microchip computer technology – manufacturers talk of "printing" the arrays. But **nucleic acid hybridization** is a more complex business than the movement of electrons in printed circuits, and proving that gene chips do what they *say* they do is not always easy.

Then, in another twist, it has also been discovered that the effects medicinal drugs produce may be modified in different ways due to many different genes, in combination with a range of environmental influences. Lots of us take several drugs at once, especially as we get older. The way our bodies handle them depends on what we eat and drink, and on other factors which may change over a lifetime. Beyond a few simple cases, the results will mostly indicate whether a drug is more or less likely to do good or harm, and doctors will still have to judge what is the best decision. They may well continue to opt for trying things out, and keeping a close watch on the patient.

There are also some worries about where such tests might lead. Some focus on the need to secure consent and confidentiality, as for any personal genetic information. Others are more specific to the drug market. **Pharmacogenetics** might reduce nasty side-effects (or "adverse reactions", as they're known in the business), but it could increase the cost of medicines as well. If tests show a drug is only suitable for a small proportion of patients, companies might hike up the price. They might even be deterred from developing such new medicines altogether.

Alternatively, early indications of who a drug might work for could cut the cost of clinical trials, and allow more compounds to get through to

Paths to breakdown

One particular set of genes that all drug companies are interested in is the set for a group of liver enzymes with the collective name of **cytochrome P450**. These evolved to neutralize toxins. There are many different P450 enzymes, which each of us produces in different combinations. They affect the breaking down of lots of important drugs, including antidepressants, anticoagulants, beta-blockers, and some common painkillers.

Roche Diagnostics, a subsidiary of the Roche drug company, markets a gene-based microarray test for variants in just one of the P450 genes. It is aimed at psychiatrists who want to know whether a patient is likely to suffer severe side effects from anti-psychotic drugs.

More generally, drug companies now routinely check how new compounds which might prove useful are metabolized by P450 enzymes. If some common variant of P450 breaks down the drug very quickly, then it will probably not be taken on into large-scale trials. Alternately, there are existing drugs, such as codeine, which are actually activated by one of the P450 enzymes. If you have enough of the enzyme, then your liver converts codeine into morphine. If you have an especially active form, you need less codeine, or the morphine will make you sleepy or even hamper your breathing. If you produce little enzyme, you just accumulate unconverted codeine, which will probably give you nausea, but no pain relief.

licensing stage by avoiding side-effects that currently stop many getting to market.

For common diseases, for which big markets beckon, this increased "specialization" of patients may not be a problem. **Rosiglitazone**, originally an anti-diabetic drug developed by Glaxo SmithKline, is now in final clinical trials for **Alzheimer's disease**, for which it may improve memory and attention in the early stages. It does not produce benefits for the whole population, but only for those who do *not* carry a particular variant of a brain protein known as **apolipoprotein E** (apoE). People who produce this particular version of apoE – apoE4 – are in fact more likely to show signs of Alzheimer's later in life. But there are still enough people who have Alzheimer's but who do *not* make the suspect brand of apoE to make it worthwhile going ahead with the trials.

This is the latest stage in a long story, in which apoE4 was first linked with an increased chance of suffering from Alzheimer's. It was marketed as the basis of a diagnostic test as long ago as 1994, but the test was shunned by most clinicians, who feared it would also be used for predicting the likely fate of younger, healthy subjects. There were further controversies about the potential use of apoE4 as a marker to help guide prescription of other anti-Alzheimer's drugs – partly because the condition is difficult to diagnose definitively until the brain is examined after death. But, although this particular story about the use of a small piece of genetic knowledge in clinical practice is not yet over, it has so far had little impact on management of this dreadful disease. **Adam Hedgecoe's** book *The Politics Of Personalised Medicine: Pharmacogenetics In The Clinic* (2004) recounts some of the reasons for this.

So what is the overall picture? All we can really say is that what happens will differ from case to case. Drug and healthcare markets are complex, and heavily regulated. This is a multi-billion dollar industry, which is finding it harder to come up with new blockbuster drugs, so there is strong pressure to discover if pharmacogenetic approaches will be any help. They have certainly encouraged the big drug companies to emphasise that conventional prescribing is a hit-and-miss game, and the importance of adverse events – both things they much preferred not to discuss until now. That in itself is an interesting change. Beyond that no-one quite knows what effect pharmacogenetics and genomics will have, but every major pharmaceutical company is thinking hard about how best to make use of the data now coming through.

Nutrigenomics: the right diet for you?

For the drug industry, genetic information offers the possibilities of differentiating diseases more precisely, and dividing patient populations into sub-groups for better tailored prescriptions. Do differences in our individual genetic make-up matter to other industries? Maybe. The commercial question is: do they offer new marketing opportunities? And the food industry is the other obvious sector in which people are thinking hard about genes.

Some of them are thinking about changing genes in crop plants, as we discuss in chapter 10. But there is also information in your genome which may reveal what will do you good on your dining table. As you will guess by now, the proponents of a new world of gene-based eating advice have a word for it: **nutrigenomics**. Making it real is still a little way off, but let's look at some of the claims, and at what might be possible.

Nutrigenomics combines nutrition and genetics in a word that can point in two directions. It stands for the study of how nutrition can affect genes and, more commonly, how our genes can affect the way we deal with foodstuffs. Generally, this is related to disease prevention. In terms of affecting genes, there are of course some foods that may harm the DNA

– and if chemicals found in food can alter gene structure, it is best to avoid them. Probably the best known example is the production of **carcinogenic nitrosamines** from nitrates in fried or barbecued meat. **Ascorbic acid** (vitamin C) is now routinely added to cured meats to help decrease formation of these chemical nasties.

Less radically, some genes may be *regulated* in ways that are affected by food. Research into the influence of the stuff we eat upon gene expression is still developing, but it is already clear that some compounds in foods can interact with particular transcription factors. For example, a whole set of genes responds directly to levels of **fatty acids**, and is expressed in quite different patterns in the liver, depending on whether a person is feasting or fasting. But in general we do not understand these things well enough yet to be able to use them as the basis for anything except more research.

The advice on offer at the moment mostly concerns the links between the way a food is processed and variations in people's genetic make-up. The thinking here is basically the same as that behind **pharmacogenetics**, but the situation is much more complicated because we are talking not just about a pure (one hopes) drug, but about foods which contain hundreds or thousands of different chemicals.

One very simple case concerns one of the success stories of the old genetics. Some newborns have an enzyme alteration that means they cannot digest the amino acid **phenylalanine**. So as they eat protein, phenylalanine – which is normally converted into another protein building block called **tyrosine** – builds up. This can cause a range of symptoms including **brain damage** as other enzymes break down the phenylalanine into other harmful derivatives. The solution was to test for the enzyme defect early (it forms part of the "**heel prick**" blood test carried out on newborns), and put babies with the altered gene on a low phenylalanine diet for life. This is not always easy to manage, but it does work provided they can stick to it as they grow up.

Another well-known example is the ability to tolerate **milk**. In most mammals the gene for processing the **double-sugar lactose** switches off once an animal is weaned, because few foods contain it. Without it, any lactose they eat cannot pass the intestinal wall. When humans began keeping cattle, there was advantage in a mutation that kept the gene in question – for the enzyme lactase – active into late childhood and onward. So people descended from cattle-rearers tend to be lactose tolerant. But you don't need a genetic test to check for this. Lactose-intolerant people who down a pint of milk feel pretty awful, and tend to suffer from bloating and diarrhoea as gut bacteria feast on the accumulated lactose.

As ever, the simple examples are instructive, but the connections between most conditions, dietary components and genes are much more complex than this. Like genes, the stuff you eat and drink has multiple effects. **Red wine** probably reduces your risk of heart disease, but almost certainly increases your chances of getting cancer. As yet, we know little about the specifics in most cases. Apart from a few disorders of **fat metabolism** such as familial hypercholesterolemia, that call for tailored diets to reduce risk of heart disease, most of the interactions are still being unravelled, and individual advice is premature.

Still, it is clear that diet has a substantial impact on chronic disease and health, and analyzing genomic data could allow the effects of food constituents to be defined more closely. This is a tall order, because foods contain many trace chemicals, as well as the bulk components such as protein or carbohydrate, and they can have lots of effects. So it is partly a matter of following how food ingredients are processed, but there is also the big challenge of investigating the effects of diverse small molecules which might, in theory, affect any stage of gene expression.

Until we have studies which generate data on all this, and tools for analyzing the results, advice will be restricted to a few well-understood genes. In time, though, it should be possible to define genetic variations which predispose people to particular diseases in ways which can be affected by modifying their diets. Early candidates include conditions like the bone disease osteoporosis, whose risk increases in women after the menopause, and is probably affected by variations in cell receptors for hormones and vitamin D.

As we say, this is in the future. But there are already companies offering personal diet advice, genetic profiling, and nutritional supplements. It is, you have to admit, a neat marketing ploy. Sell folk the test. Then give them a result which makes them come back for products you offer.

An insight into how this works comes from a **US Government Accountability Office** (GAO) investigation in 2006. The GAO bought tests for, supposedly, fourteen people of different ages, weights and lifestyles from four different websites. In fact all the DNA samples came from one man and one unrelated woman. The test results sent back claimed the people GAO invented were at risk of various conditions, including osteoporosis, cancer, type 2 diabetes, high blood pressure, "brain ageing", and reduced ability to clear toxins. The results came with advice that usually began by reinforcing standard healthy-living recommendations (eat sensibly, exercise, stop smoking). But two of the companies followed through with recommendations for dietary supplements supposedly tailored to the DNA of the subjects. The GAO reports that these were ordinary multivitamins, which anyone can buy for a few tens of dollars a year. The supposedly "nutrigenomic" packages would have cost over $1000 a year. One even claimed to aid DNA repair. Overall, the US investigation concluded that the companies' claims were "both medically unproven and meaningless". The tests, from the companies Suracell, Genelex and Sciona, were about as much use as genetic horoscopes, the survey director concluded.

This was pretty much the same verdict that the UK's **Human Genetics Commission** delivered when it looked into nutrigenomic tests. It led to one of the four companies taking its test off the market in Britain – and shifting their marketing to the US. So the best advice, for now, on both sides of the Atlantic is: take vitamins by all means, but save your money by sticking to the generic tablets.

While we wait for a more convincing analysis of individual food, gene and health interactions, however, we can expect to see other developments that point the way to future marketing strategies based on genomic science. The salt industry, for instance, is worried about the pressure to cut the salt content of processed foods, which is unhealthy for everyone. But they would still like to patent genes that might be candidates for identifying "**salt-sensitive**" people within the general population. This would have some immediate advantages for the industry. It could argue that only *some* people need to avoid high-salt foods, and leave that demographic alone. Then, just as diabetics can buy low-sugar cookies and sweets, there would be plenty of marketing opportunities for higher-cost alternatives for the salt-sensitive.

Governments will inevitably be interested in the potential of dietary advice related to genetic information. It would be tricky to implement, as it is already hard enough getting across general, good-sense diet and health advice that applies more or less to everyone – introducing complications involving individual genetic variation runs the risk of obscuring the messages we already hear about, say, reducing fat or eating more fruit. But reliable, genetic dietary advice, based on reputable science, would fit public health targets and be good for health budgets – food is, after all, cheaper than drugs, and what's more people expect to pay for it themselves.

If such advice amounted to the simple avoidance of some foods, it would be simple enough to help people to do so. But if **dietary supplements** were called for, there might be problems with targeting the groups who need information, and perhaps with over-use of them by the enthusiastic. None of this is entirely new of course. But when you consider that genomic data on individuals will inevitably presage crops modified to produce key chemicals – "**nutraceuticals**" – you can easily imagine a world in which the distinction between food and drugs begins to blur, with all the regulatory tangles that implies.

For now this remains largely speculative. You can follow developments at the website of the recently founded **European Nutrigenomics Organisation**, which was set up in 2004, but is still very much a work in progress. See www.nugo.org/everyone

Over-the-counter genetic testing

At the moment genetic testing is mostly available through doctors, in both public and private medical care systems. As the technology becomes cheaper and more widespread, this may change. Interpreting a read-out of multiple **alleles**, or even your whole genome, will likely remain a matter for the professionals. And more sophisticated profiling of individual cancers, say, will remain in the hands of specialists. But we might allow people to check for single-gene variants in the same way that you can go to your chemist or drugstore and come back with a home cholesterol test. Both give information that may relate to risk of disease. So are we on the way to over-the-counter – or over-the-Internet – gene tests?

Medical professionals tend to argue that **direct-to-consumer** (DTC) gene tests will need strict regulation. They argue that the tests must be reliable, and require proper analysis, careful interpretation and explanation – by doctors or genetic counsellors. This could be seen as protecting professional interests, but there is growing concern that without proper explanation people could misinterpret results of genetic tests. They could either join the "worried well", living in fear of developing a disease that they have little risk of developing, or embark on costly medical treatment which they do not need.

Striking the right balance here may not be easy. Governments increasingly urge people to take responsibility for their own health. While gene tests could be manna to hypochondriacs, they could also contribute to a personalized approach to preventative medicine. But professional bodies remain cautious. The Board of Directors of the **American College of Genetic Medicine** said in 2004 that they condemned the sale of gene tests via the Internet as they could harm physical health – through "inappropriate utilisation" – and peace of mind. They believe that gene tests are special, and must be provided only when genetic counselling is available.

Some commercial providers clearly disagree. **Myriad Genetics**, who hold most of the rights to testing for the two breast-cancer genes, **BRCA1** and **BRCA2** (see p.70), were also the first company to market direct-to-consumer genetic tests. In 1996 they released a DTC test for BRCA1, but they withdrew it after criticism that it was "reckless". In 2002 Myriad launched a new DTC campaign. This included extensive media advertising. However, research showed that although this increased the public's awareness of the availability of the tests, it did little to reach the small percentage of women for whom the tests are relevant. Myriad were cautioned by an advertising watchdog for "failing to accurately convey risk information" in their advertising campaign for Reanalysis. Myriad currently offers testing and results through doctors, and not direct.

At the time of writing the vast majority of other commercial testing services offer genetic fingerprinting – for paternity, maternity or sibling testing, ancestry and forensic uses. But a few companies do now offer "home-testing" kits for a variety of diseases. **DNA Direct**, for example, will provide tests for alpha-1 antitypic deficiency, haemochromatosis, cystic fibrosis, breast and ovarian cancer and blood clotting disorders. New tests for drug response will be "available soon". Their website (www.dnadirect.com) claims they are recommended by *Good Housekeeping* magazine. More convincingly, they meet the recent standards proposed by the US Food and Drug Administration and Center for Disease Control. The company maintains a highly informative blog about genes and testing at talk.dnadirect.com Other companies offer tests for risks of heart disease, Alzheimer's and diabetes. There are also tests available which claim to shed light on whether people are at high risk of cancer if they smoke, or claim that they can be used as a basis for dietary advice. (We discuss the latter under nutrigenomics: see p.250).

Most of the tests work this way: you request a kit which is posted to your home address and you return it to the company with a cheek swab DNA sample. Tests are done, and about seven to ten days later you receive notification that your results are ready, and you can pick them up on a password-protected server online. Price range from a few hundred to over a thousand dollars. There may or may not be a counsellor available to help you make sense of the information. If you do contemplate getting a genetic test done, the latest advice from the **US Food and Drug Administration** and the **Center for Disease Control** is hard to improve on. The details quickly get complex, which is why their recommendation to talk things through with a qualified counsellor needs to be taken seriously.

Government regulation for these tests is still being worked out. The **UK Human Genetics Commission**'s most recent recommendation was that DTC tests should be strictly controlled, but not banned. One concern they highlight is that people would begin the testing on their own initiative, then fill up the doctors' surgeries seeking further advice or even more tests. They also registered concerns about testing children without their consent. The companies, however, argue that they are helping consumers by offering them tests in the privacy of their own home rather than through their doctors or health insurance providers – not least because there are serious worries that knowing about genetic defects will raise insurance premiums or make people less employable. The **US Genetic Information Nondiscrimination Act** of 2005, makes it illegal to "fail or refuse to hire or to discharge any employee, or otherwise to discriminate against any employee ... because of genetic information ... (or information about a request for or the receipt of genetic services by such employee or family member of such employee)." But it is easy to see why some people would still prefer to conduct such tests privately. As with other aspects of genetic testing, the regime that will ultimately be put in place for tests marketed directly to the public will depend on the extent to which information is judged special just because it is genetic information. Current indications are that most of those involved in the discussion believe that in some respects it is.

Future/better humans?

The longer-term possibilities of new genetic technologies are harder to get a handle on than the ones we have discussed so far. Prediction is a game anyone can play, but it quickly shades into wishful thinking, or plain science fiction. But here we stick with what some consider, or what may become, real possibilities.

It may be helpful to divide the technologies that could be on the way into two categories. There will be new technologies used to read, store and analyse information from **living tissue samples** of all kinds. Then there will be others for the trickier business of manipulating **DNA inside organisms**. The former are concerned with creating external readouts of the interior states of cells. Most of us have cells to spare, so we will not mind if the rather small samples needed are destroyed in the process. But the second set of technologies effectively involves *changing* what is inside the cell to fit ideas about a new, improved version of what was there before. Granted, this will probably still happen at the level of the information written in DNA code. But it has to be done in a way that leaves the rest of the cell intact; as there appears to be an almost unlimited number of ways for cells to go wrong, it will remain much harder to do than extracting the information.

It seems a safe bet that it will get easier and easier to read genetic information. Just as you can plot the increasing speed and falling costs of computers over the past several decades, so the cost of **DNA sequencing** is coming down fast. The cumulative cost of the publicly funded Human Genome Project was around $3 billion, but as techniques improved the main part of the sequence cost between $300–$500 million, depending on when you count from. A complete human genome can now probably be sequenced for around $20 million. But genetic futurists are looking toward the "thousand dollar genome", a point when everyone – well, every afflu-ent Westerner – can have their own complete sequence, if they want.

How far off is that? Well, the **US National Institutes of Health** put up $70 million in grant money in 2004 to encourage genetic futurists to come up with ways of delivering a $100,000 genome by 2009 and a $1,000 genome five years after that. They found plenty of takers. The target is ambitious – remember there are 6 billion base-pairs in a single human genome (in two sets of chromosomes). But it seems likely that application of miniaturization, automation, and plain old ingenuity will bring success. There are private sector efforts, too, and a good many ideas to try. They include improvements in the chemical routines already used, new physi-cal techniques which "read" DNA, and new set-ups which use part of the cell's own DNA-processing machinery.

The idea of the "thousand dollar genome" got a lot of press in the sum-mer of 2006. But it probably is not something to get too excited about. If we really are offered personal genome readouts, they will have to be more accurate than those of the current DNA sequencers to be of much use – which will be a tall order. If that were achievable, they will consist of a mass of information, and it will be sensible to keep this under some protection for confidentiality – but this need be no more complex than measures already adopted for other medical records. Genes and DNA are significant, but not much more significant than other things you can know about a person. It is also worth bearing in mind that it would be a static readout, which really needs to be interpeted in the light of the other "omic" data we talked about in chapter four. So moving into the area of personalized and predictive medicine, which some still foresee in the near future, will depend on establishing **systems biology** as a medically useful set of theories.

Enhancing our genes

Post-human futures?

As with the advent of systems biology, there are good reasons to be sceptical about the predicted availability of gene therapy any time soon. There are layers of complexity in coordination and regulation of genes which are still being uncovered in the wake of the **Human Genome Project**. So plugging in new genes, or altering existing ones, to achieve an effect specified in advance, may prove trickier than people used to suppose.

It is also true that most countries which have any policy at all restrict experiments in gene therapy to body cells, where they would not be inherited. So you might argue that the issues of **germ-line therapy** – the altering of eggs or sperm – are all merely academic at the moment. But it is still worth thinking what the consequences might be if it were possible, and about the arguments for and against.

Altering future generations

The germ-line/somatic cell distinction avoids a number of issues that trouble many people about messing with sperm or eggs. (Somatic cell is the biological term for body cells.) Obviously, the changes would be passed on, so future generations would bear the cost, as well as reap the benefits (though presumably if the technology actually worked, they would be able to change themselves back!). Getting the technology to work at all may involve experiments on embryos, which in itself currently rules it out for some.

But assume these objections do not completely rule out the project. Perhaps we can perfect it in model organisms to a point where everyone is convinced it is safe and reliable. And there are sure ways in which inserted genes might be disabled so that they were not inherited by the recipient's descendents. After all, why would you want to alter the germ-line?

One reason might be simple medical "efficiency". Why leave a child with debilitating and life-threatening cystic fibrosis (even if there was a somatic therapy which might cure it) if you could alter germ cells so that cystic fibrosis did not arise in the first place? Then there are arguments that propose going *beyond* correcting a genetic defect. One might imagine altering genes to increase resistance to particular diseases. Having a built-in resistance to the AIDS virus, for example, might well be considered a legitimate improvement at genetic level. But these are the relatively "straightforward" propositions. Others are more philosophically interesting because they are simultaneously regarded as reasons to allow germ-line alteration by some people, but seen by others as reasons to view it warily. And they inevitably take the discussion beyond the realms of medicine.

In fact, even medicine moves beyond the realms of medicine, as it were. After the "somatic versus germ-line" distinction, another one which is often drawn is between **remediation** and **enhancement**. But unlike the first, this one does not really hold up under examination. The idea is that we might be able to regulate the uses of gene technologies so that they just fixed problems, cured or prevented diseases or, more formally, restored some agreed level of normal species function. Going beyond that would not be legally permitted. But there are plenty of cases where it is hard to draw the line between treatment and enhancement. A great deal of medical research, to take just one example, is targeted at diseases related to old age. But there is no real difference between the biology you would need to understand to be able to fix these, and what you would need to know to actually extend the human lifespan.

At this point, then, opinion divides rather starkly between advocates and critics of the idea of human enhancement. A good sample can be found in the books published in 2002 by futurist **Gregory Stock** and political scientist **Francis Fukuyama**. Stock, in *Redesigning Humans* suggests that it will not be possible to stop enhancement, and that it should be a matter of individual choice anyway. If there are adverse consequences of the sum of individual decision, we can respond in time to correct them. But in general, we should welcome the chance to improve the human lot by altering ourselves for the better. Parents want the best for their children, and already try to improve their lot through education. Why bar them from achieving the same ends through different and maybe more effective means just because it involves genetic change?

Like Stock, current participants in these debates are looking forward to a time, perhaps not too far off, when humans use technology to transcend their present condition. Eventually, they might even become what are referred to as "**post-humans**". The vicissitudes of evolution have left us with bodies and minds which are badly in need of improvement. Let's give evolution a helping hand and make whatever improvements we can, goes the argument. This transhumanist vision, as its advocates call it, embraces genetic alteration as one of a set of transformative technologies, along with super-powerful computers, artificial intelligence and nanotechnology. Together, they signal the advent of the "singularity", the point at which unlimited technological transformation of humans and their environment becomes self-sustaining.

Assuming that that particular future remains science fiction (of which more later), let's stick with genetics and consider the downside of altering humans. **Francis Fukuyama**, who has debated with Stock on platforms round the world, argues in *Our Posthuman Future* that changing the human genetic make-up is deeply undesirable because altering the essence of humanity, as he sees it, would dissolve notions of **human rights**. Call it the "keep humans human" position. It rests on a notion of a fixed human nature, to which the best of all possible political systems – liberal democracy – is perfectly adapted. As he sees it: "Human nature is the sum of the behaviour and characteristics that are typical of the human species, arising from genetic rather than environmental factors ... Every member of the human species possesses a genetic endowment that allows him or her to become a whole human being, an endowment that distinguishes a human in essence from other types of creatures."

This is how we are, says Fukuyama, and this is how we should stay. It is the only basis for defending human rights, and any deviation from the human essence would compromise those rights. If some people were enhanced, they would in some sense no longer be human. How would we know how to treat them? How would they know how to treat us? Fukuyama hints that the results would be apocalyptic. Our ideas of justice and equality are grounded in an idea of human rights which derives from the conviction that there are qualities all humans have in common, he argues. Once, women and "inferior" races were not granted these rights, and it has taken a long struggle to show that they must be fully included in any just political and legal system. Lose that assumption by creating a world in which some people are more equal genetically than others, says

Fukuyama, and all bets are off. The likely result is a "genetic arms race", between neighbours, classes, even nations. Who knows where it would end?

Who is right? Damned if we know. It seems hard to agree that humans as we know them are beyond any improvement. And life extension, for example, if it were ever on offer, would be a technology likely to meet overwhelming demand. Whether that is a good thing, and whether we could handle the new problems of justice, equity, or rights which serious genetic enhancement might pose, is more difficult to say. One group of academic bio-philosophers in the US has argued that the possibility of designed genetic change will demand a radical rethinking of notions of justice, but from a very different standpoint. In their demanding book *From Chance To Choice*, four bioethicists point out that existing theories of justice try to deal with the fact that people's individual endowments at birth need not be equal – hence notions like equality of opportunity, for instance. But if we had the ability to alter those endowments, perhaps justice would oblige us to? Again, there are no simple answers. But it is at least possible that these will become real questions, rather than largely academic ones, before the twenty-first century is out.

A brief history of the post-human

It may be turning into one of the big twenty-first century questions, but the idea of human enhancement has a longer history, of course. Ever since the idea of evolution took hold, people have imagined future humans as different. If species can change, then perhaps our descendants will be fabulously physically endowed. Or maybe they will be degenerates, a pale shadow of the gracious creatures we see around us now.

This was a quite new take on history. The idea that the future would be different from the past took hold during the Enlightenment, and seemed confirmed by the scientific and industrial revolutions of the seventeenth and eighteenth centuries. This tended to go along with a vision of progress. The best that we could hope for in earthly life was no longer a return to some Arcadian past, but a future of man-made perfection. Such dreams focused on social, technological and political change. But with evolution came the thought that *biological* change was also possible. Whether this would still be progress was harder to figure out. Fiction writers explored a range of possibilities – largely intuitively they have come up with just about all of the range of logical possibilities for how change, and rates of change, might happen over future evolutionary time. If you start from a prognosis of no change, the opposite extreme would be indefinite, perhaps infinite progression. Or we might witness advances to new levels, then stabilization; or advance followed by catastrophic collapse. On the other side of the coin, if there is some kind of *regress* – back to the cave or the primeval slime – this might be followed by a period of little change, or by some sort of renewal. Enough of the logic: here are some examples.

H.G. Wells wrote two great Darwinian fables. The first was *The Island Of Dr Moreau*. The second, *The Time Machine*, was published two years later. It had one unforgettable vision of those distant descendants of the Victorian class system: the feeble-minded Eloi; and the sinister, subterranean Morlocks. The time-traveller's first glimpse of the latter is of "a queer little ape-like figure, its head held down in a peculiar manner, running across the sunlit space." It "was a dull white, and had strange large greyish-red eyes" with "flaxen hair on its head and down its back". But it runs away from him too quckly for him to say "whether it ran on all-fours, or only with its forearms held very low".

This is clearly Wells' vision of a degenerate human, an impression reinforced by the subsequent revelation that the Morlocks use Eloi for food. Both are victims of the success of their technology; then of its failure. Wells was writing before Mendel's genetics was rediscovered, so there is no actual genetic technology involved in his imaginings. The returned time-traveller explains:"the Upper-world man had drifted towards his feeble prettiness, and the Under-world to mere mechanical industry. But that perfect state had lacked one thing even for mechanical perfection – absolute permanency. Apparently as time went on, the feeding of the Under-world, however it was effected, had become disjointed. Mother Necessity, who had been staved off for a few thousand years, came back again, and she began below. The Under-world being in contact with machinery, which, however perfect, still needs some little thought outside habit, had probably retained perforce rather more initiative, if less of every other human character, than the Upper. And when other meat failed them, they turned to what old habit had hitherto forbidden."

The changes in *The Time Machine* are simply the result of continuing selection under changed conditions. This is the territory of **eugenics** (see p.154). And it is not long before eugenics is beefed up by artificial aids, though still as an aid to selection rather than through actual engineering of new variants. In English geneticist **J.B.S. Haldane**'s famous speculative essay *Daedalus* of 1923, for example, **ectogenesis** – test-tube baby technology – is used as a technical fix to stave off degeneration. "The small proportion of men and women who are selected as ancestors for the next

H.G. Wells, author of the Darwinian fables *The Island Of Dr Moreau* and *The Time Machine*

generation are so undoubtedly superior to the average that the advance in each generation in any single respect, from the increased output of first-class music to the decreased convictions for theft, is very startling. Had it not been for ectogenesis there can be little doubt that civilization would have collapsed within a measurable time owing to the greater fertility of the less desirable members of the population." Haldane was one of a group of writers from the 1920s and 1930s, most but not all of them Englishmen, who created many of our lasting images of genetic futures.

Everyone has heard of **Aldous Huxley**'s *Brave New World* of 1932 – not so much a vision of human evolution, as of evolution at an end. In the novel, technology has brought enormous change. But that change is over. The rulers' attitude to technological innovation is closer to that of Imperial China than to the capitalist culture Huxley was satirising. As the novel's chief technocrat, Mustapha Mond, puts it: "We have our stability to think of. We don't want change. Every change is a menace to stability. That's another reason why we're so chary of applying new inventions. Every new discovery in pure science is potentially subversive; even science must sometimes be treated as a possible enemy. Yes, even science."

The key to stability, he emphasizes, is Bokanovsky's process, a kind of cloning, ensuring a reliable supply of compliant gammas and deltas to do the (unnecessary) menial work. Although often remembered as a picture of unlimited genetic meddling, *Brave New World* utlimately shows progress halted.

You can see that more clearly in the breathtakingly expansive vision of the Marxist crystallographer and technological visionary **J.D. Bernal's** *The World, The Flesh And The Devil* of 1929. He sees future mankind completely transformed, albeit not through genetics. Instead, his future humans turn to electro-mechanical enhancement, and fashion a kind of super-cyborg. As he wrote: "The new man must appear to those who have not contemplated him before as a strange, monstrous and inhuman creature, but he is only the logical outcome of the type of humanity that exists at present .. Although it is possible that man has far to go before his inherent physiological and psychological make-up becomes the limiting factor to his development, this must happen sooner or later, and it is then that the mechanized man will begin to show a definite advantage. Normal man is an evolutionary dead end; mechanical man, apparently a break in organic evolution, is actually more in the true tradition of a further evolution." Bernal gave us both the first instance of future man as a kind of brain in a vat, and a stark reminder that more than one kind of evolution was now possible, and that there might be crucial trade-offs between them.

Olaf Stapledon took up all these possibilities and more in his astonishing novel, *Last And First Men*, published in 1930. This, along with his later *Star Maker*, is the grandest evolutionary story of all, and a huge influence on later science fiction. *Last And First Men* runs the whole catalogue of possibilities for future humans. Stapledon relates the history of a succession of species descended from us – the "First Men". The first new departure comes about through the normal working of evolution. "It was some ten million years after the Patagonian disaster that the first elements of a new human species appeared, in an epidemic of biological variations, many of which were extremely valuable. Upon this raw material the new and stimulating environment worked for some hundred thousand years until at last there appeared the Second Men".

These second men then reconstruct themselves, in different ways at different times, over a period of some two billion years. They begin with genetic alteration, move to the electro-mechanical maintenance of abstracted giant brains, then opt for bodies again, to recover their essential humanity. In some ways, there is little to add after this sweeping narrative, although *Last And First Men* is confined to the solar system, whereas Stapledon's *Star Maker* went on to open the canvas to include other stars and galaxies.

More earthbound writers, however, continued to develop ideas of directed improvement of the human species. Pioneer geneticist and life-long eugenicist **Hermann Muller** did this in his would-be prophetic tract of the 1930s, *Out Of The Night*. "In time to come, the best thought of the race will necessarily be focused on the problems of evolution – not of the evolution gone by, but of the evolution still to come – and of the working out of genetic methods, eugenic ideals, yes, on the invention of new characteristics, organs, and biological systems that will work out to further the interests, the happiness, the glory of the god-like beings whose meager foreshadowing we present ailing creatures are". Evidently he couldn't wait. Later in the twentieth century, of course, genetic futures were discussed more often, and more urgently. For quite a while after the discovery of the DNA double helix, genetic engineering (a phrase popularised in the 1960s) was easier to imagine than the cyborg vision of people like Bernal. In his landmark book of 1968, British journalist **Gordon Rattray Taylor**'s *The Biological Time Bomb*, Taylor postulated that: "To judge from what the scientists themselves are saying, the most serious of all the human problems created by biological research is constituted by man's imminent power to interfere in the processes of heredity to alter the genetic structure of his own species".

That idea is now pretty commonplace, though not always seen as a problem. Some writers – such as British biologist **Adrian Woolfson** – simply take it for granted that it will happen. In his intriguing book from 2000 *Life Without Genes*, he puts it like this: "When we have charted the genetic landscapes that have been explored by natural evolutionary processes and are in a position to fully appreciate the nature of the mechanisms responsible for generating and modifying living things, life will enter a new realm of history. It will no longer lie in the exclusive and capricious historical domain of chance and natural selection. It will instead be possible to design and construct new living things using ahistorical processes, in much the same way that we currently design and construct motor cars, traffic lights, helicopters and vacuum cleaners".

This takes the engineering metaphor to the max. Woolfson does not discuss what this means in much detail, but lots of other people do. That takes us back to the two main camps represented by **Francis Fukuyama** and **Gregory Stock**, but gives a different slant on where they are coming from. For Fukuyama, stability is the aim – and Huxley's warning should

Author of the dystopian *Brave New World*, Aldous Huxley

be taken seriously. As we have said, this is a misunderstanding of Huxley's book, but plenty of other writers use it in the same way.

Stock, however, believes that it is our nature to experiment. He suggests that "to turn away from germ line selection and modification without even exploring them would be to deny our essential nature and perhaps our destiny." Striking, that sudden appearance of "destiny", after chapters of relatively sober exposition and argument. For Stock is gripped by the grand evolutionary narrative. His inspiration comes not from Huxley, but from the likes of Bernal, Muller and Stapledon. As he goes on to say: "The project of humanity's self-evolution is the ultimate embodiment of our science and ourselves as a cosmic instrument in our ongoing emergence." Gosh.

What are we to make of all this? How, for example, does it relate to the details of specific bits of genetic technology we can use now, and to the nitty-gritty of policy debates, laws and regulations which you can find elsewhere in this book? Well, it at least shows that there is another debate, on a grander scale, which goes on alongside these discussions, and which has been around for some time. And it demonstrates that it really matters where people's allegiances in that larger debate lie – whether they acknowledge them or not. It makes a difference whether you believe, in the end, that humans ought to be left to make the best of life pretty much as they are now, or whether they should take their evolution into their own hands to "realise their cosmic destiny". The difference in the twenty-first century is that we (that is, you) may actually get to choose.

PART 6
GENES, POLITICS AND CULTURE

What will happen next?

And who decides?

No one knows where knowledge of genes will lead us. There have been plenty of predictions, as we have seen. And reviewing the progress of the last fifty years, it's clear that at least some of them will come true. The potential consequences demand our attention. Genetic technologies touch on intimate concerns: on life, death and identity. They could transform sectors of industry and agriculture, and perhaps one day even alter future generations of humans. But who do these technologies belong to? Is anyone really in control?

While governments continue to pay for some of the work conducted in the field of genetics, especially in terms of basic research and health care, much lies in the hands of private companies. Safety rules and other regulations – though they differ from country to country – have to be enforced. Of course, governments recognize that these technologies are important. They often want to promote them. But there are mechanisms in place to alert people to the issues they raise, and to decide how they should be regulated, if regulation is necessary. As in other areas of science and technology, regulators have to balance benefits against harms – to patients, consumers, the environment and to national economies.

The experts

There is a forest of official advisory committees and regulatory bodies – national, international and regional – on various aspects of **genetics**, **biotechnology** and new **reproductive technologies**. Their status and terms of reference vary. When you factor in departments of state, and other authorities whose interests impinge on biotechnology, then the business of finding out who to ask about what – or who might be interested in your opinion anyway– becomes pretty complicated.

It used to be annoyingly difficult to keep track of all these bodies, to say nothing of the pressure groups and NGOs active around gene technologies. New controversies tend to spawn new committees. For example, when the use of tissue from babies and children who had died at Alder Hey Children's Hospital in Liverpool was made public in 1999, and it was found to have gone ahead without parents' permission, there was a national outcry in the UK's media. After an official inquiry, the British government moved to set up the **Human Tissue Authority** to regulate scientists' use of human samples. It became one of ten "arm's length bodies", as the UK civil service calls them, answering to the Department of Health. But it was decided almost immediately that it would be merged with the long-standing **Human Fertilization and Embryology Authority**, albeit not until 2008/9.

The Internet has made it much easier to keep track of all the various "Authorities". Along with a new – in the UK, at least – commitment to openness, it has encouraged all these bodies to maintain their own websites. As well as reports, minutes and notices of meetings and consultations, most of them have good links to other relevant sites. If you know one or two of the important ones, you can usually find the rest.

Bodies and bioethics

Governments must hate genetic technologies. Not for their economic and health benefits, of course, but for the political problems they cause. They keep throwing up challenges to accepted ways of doing things and established categories, which make them hard to handle. Uses of genetic technologies that could be problematic, or which require regulation – *most* of them, then – generally fall under the heading of what is nowadays known as **bioethics**. This is a relatively new field, born out of concern about the potential abuses of medical techniques following the range of atrocious practices conducted by scientists during World War II – and the protection of human subjects in research. Now the term covers lots of other things, from the supply of drugs to developing countries to transplant surgery.

Most countries now have national bioethics committees. They will typically have a standing committee, and will form sub-groups to report on particular issues and to advise legislators, or anyone else, on how to think about an issue, or what should be done. They are usually comprised of a cross-section of specialists, both medical and otherwise – not just scientists. There will typically be religious representatives, and an academic bioethicist or two, who may be a specialist or a philosopher with other interests too. And increasingly there will be one or two members who represent lay people or consumer interests. In practice, it can sometimes be hard to figure out who some of the members of bioethics committees do in fact represent. Add a sprinkling of doctors, lawyers, and perhaps a biotech executive, and you have your bioethics committee.

The UK does not have an official committee charged with general responsibility for biological and ethical issues. As a result, the **Nuffield Council on Bioethics**, sponsored by the Nuffield Foundation, was set up. It now operates along very much the same lines as national bioethics advisory committees in other countries, with support from the big medical research charity **The Wellcome Trust** and the **Medical Research Council**. It does not have a direct line into government, however, and the fact that most of its support comes from two big research funders does not exactly make it look independent. In practice, its reports are authoritative and are widely read, although it is hard to know what influence on policy they have. You can find them online at www.nuffieldbioethics.org

Since it began work in 1991, the Nuffield Council has produced reports on the ethics of research involving animals, research related to health care in developing countries, **pharmacogenetics**, genetics and human behaviour, patenting DNA, genetically modified crops, stem cell therapy, mental disorders and genetics, animal to human transplants, human tissue and genetic screening.

In the US, the **President's Council on Bioethics** covers a similar range of issues. It was established by President **George W. Bush** to succeed the earlier National Bioethics Advisory Commission. Mostly, the Bush administration was content to purge existing committees and advisory panels of anyone with the remotest liberal tendencies, but it was decided that, for bioethics, a more complete revamp was needed. Controversially, the new Council's first chairman was **Leon Kass**, a long-standing critic of practically every new biological technology, from IVF to cloning.

Unusually, the President's order creating the Council said that "the Council shall be guided by the need to articulate fully the complex and often competing moral positions on any given issue, rather than by an overriding concern to find consensus". So their report on regulating new reproductive technologies, *Reproduction And Responsibility* from 2004, for example, closed with a series of personal statements from members about how they viewed the findings and recommendations.

In this vein, it has so far produced reports on human cloning, human enhancement, stem-cell research, regulation of new reproductive technologies, and care of the ageing. The enhancement report of 2003, *Beyond Therapy: Biotechnology And The Pursuit Of Happiness*, was a particularly weighty production, running to book length and reviewing many fascinating aspects of the effort to be "better than well". You can read it, and their other reports, at www.bioethics.gov which also has links to earlier US committees and their reports.

Most other countries also have national bioethics advisory committees, commissions, and councils (or whatever their favoured term is). They all have wider remits than just gene technologies, but most spend a good deal of their time on genetics and related issues. The main ones in the English-speaking world are:

Australian Health Ethics Committee
www.nhmrc.gov.au/about/committees/ahec/index.htm

Canadian National Council on Ethics in Human Research
www.ncehr-cnerh.org

Irish Council for Bioethics
www.bioethics.ie

Toi te taiao: New Zealand's Bioethics Council
www.bioethics.org.nz

The Nuffield Council's website has a list which covers many other countries. Internationally, the **Council of Europe** has a **Steering Committee on Bioethics** (CDBI) (www.coe.int/bioethics). The **OECD** has an Internal Coordination Group for Biotechnology (ICGB). Their website (www.oecd.org/biotechnology), is good for statistics and for analyses of what they call the "bioeconomy". UNESCO has an **International Bioethics Committee** (www.unesco.org/ibc).

But beneath the level of national bioethics committees, there will be aspects of genetics being assessed and ruled upon through further departments, committees or agencies – in the administration of health care, food, industrial health and safety and environmental protection, for example. (The initial lead in the **Human Genome Project** was in fact taken by the US Department of Energy.) Their general attitudes towards the implementation of genetics may vary. Here are some of the significant bodies involved, first in the UK, then the USA.

The limits of bioethics?

The rise of bioethics is in many ways a logical response to the burgeoning of biomedical technologies, and of concerns about their application. But bioethics has developed a particular style of considering what should be allowed. The American sociologist **John Evans** has argued at length that the dominance of the bioethicists has led to what he calls a "thinning" of debate about key issues, especially human genetic engineering. When theological influence was stronger, he believes, there was more discussion about ultimate ends – the purpose of life, if you like. Come the new profession of bioethicists, and discussion became narrower, focusing above all on the decisions of individuals and not on broader social goals or concerns, or upon big questions such as the kind of people we should try and become.

Evans identifies a key moment in the US as the formation in 1973 of the **National Commission for the Protection of Human Subjects of Biomedical and Behavioural Research**. The new commission was supposed to find some general principles which it could apply across a wide range of research. It settled upon "respect for persons, beneficence and justice". These three were the principles they took to be generally acceptable, though without consulting the wider public. They were then meant to be safeguarded through the legal condition of informed consent (respect for persons), risk/benefit assessment (beneficence) and fair procedures for selecting research subjects (justice).

Evans says that, while this pioneering commission was at the time concerned with research on human subjects, its rules have carried over into broader bioethical discussions. He concludes this has left bioethics poorly equipped to deal with the aspects of human genetic engineering that really matter – such as whether a society ought to be doing it at all. You can read more in his book *Playing God? Human Genetic Engineering And The Rationalization Of Public Bioethical Debate* (2002).

In the UK

The **Human Genetics Commission** (HGC) is the UK Government's advisory body on new developments in human genetics and how they impact upon individuals' lives. Its remit includes social, ethical and legal issues, but it has no regulatory powers of its own. The commission says that one of its key roles is to promote debate and to listen to what the public and its stakeholders have to say. The current chair is Baroness **Helena Kennedy** QC and the commission has 24 members including experts in genetics, ethics, law and consumer affairs. It also has a consultative panel of people who have direct experience of living with genetic disorders.

Sub-groups monitor special topics, including identity testing, genetic discrimination, intellectual property, databases and genetic services.

Reports and consultations on these and other topics are available to download from the HGC's website: www.hgc.gov.uk

The UK Department of Health also has a **Gene Therapy Advisory Committee** (GTAC), which was set up in 1993 to assess gene therapy proposals on ethical grounds. GTAC approval must be obtained before **somatic cell gene therapy** (ie on any cell other than the sperm or egg cells) or gene transfer research is conducted on human subjects. The GTAC will not, at present, consider proposals for **germ-line cell** (egg or sperm) gene therapy. Its main concern is whether proposals for gene therapy studies meet ethical and safety criteria for research on humans. In 2005, the committee considered fifteen applications from researchers, and approved twelve. Gene therapy trials count as clinical trials, and are regulated by the **Medicine and Healthcare Products Regulatory Agency** (www.mhra.gov.uk), which takes advice from GTAC. For more information, see www.advisorybodies.doh.gov.uk/genetics/gtac

The **Genetics and Insurance Committe** is yet another UK special committee set up to deal with a controversial issue, having been established to advise the Department of Health on whether genetic tests could be used by insurance companies. It was the government's response to a Parliamentary select committee report in 1999, when there was widespread concern that people who had taken a genetic test would have to tell their insurers. But the fifteen-member committee has not exactly been busy. It has only approved one application, in 2000. This was for insurance companies to take into account the results of **Huntington's disease** tests for life-insurance policies worth over £500,000. At the time of writing there were no further applications under consideration, and none in prospect. The Association of British Insurers told the committee it would not be submitting any applications to use predictive tests, including for breast cancer, during 2006 and 2007. See www.advisorybodies.doh.gov.uk/genetics/gaic/index for more information. The **Genetics Commissioning Advisory Group** (GenCAG) is not a very public group, although it does publish minutes (see www.dh.gov.uk/PolicyAndGuidance/HealthAndSocialCareTopics/Genetics). But it's a very important one for the development of genetic services. It takes a "strategic national overview" of genetics in health care delivery. The aim is to provide advice to commissioners of genetics services – that is, local health-service managers - to enable them to provide appropriate services for NHS patients and their families.

The UK's **Human Fertilization and Embryology Authority** (www.hfea.gov.uk) was one of the earliest statutory bodies set up to oversee a new set of biological technologies. It was the outcome of a lengthy debate after the birth of the first "test tube" baby, **Louise Brown**, in 1978. Following the recommendation of a high-powered committee chaired by the philosopher **Mary Warnock** in 1984, the HFEA now oversees fertility clinics using established technology and research involving human embryos. It was widely regarded as a model agency, and its work has grown over the years as *in vitro* fertilization and associated technologies have become the basis of a multi-million pound industry, and the research has expanded into new areas.

Although the HFEA is primarily concerned with human embryos, rather than genetics, in practice it is the most important decision-making body when it comes to the use of genetic tests carried out on IVF embryos – **pre-implantation genetic diagnosis**. PGD can only be carried out in licensed fertility clinics, and is approved on a case-by-case basis. This has lead to the curious situation in which PGD is subject to far more regulation than any other form of medical test in the UK – including prenatal genetic tests carried out on fetuses. The HFEA is often at the centre of controversy, either about the precise conditions under which people are allowed to try and have children, or about what kinds of research or testing are allowed. It also now regulates research using human embryonic stem (ES) cells, which fall under the general condition that licensed work with embryos (both those left over from fertility treatment and those created specifically for research) is acceptable as long as they are not maintained in the lab for more than fourteen days – the developmental stage Warnock dubbed the "pre-embryo". As in other countries, anti-abortionists oppose this kind of work and indeed any form of IVF procedure, since it often produces more embryos than are required for treatment – but British legislation continues to permit it.

In fact the Human Fertilisation and Embryology (Research Purposes) Regulations 2001 widened the permitted experimental uses of human embryos beyond those originally specified in 1990, specifically in order to permit research aimed at deriving ES cells and exploring their therapeutic potential. Previously, research on embryos was only permitted if its aim was to shed light on infertility, or to develop new fertility treatments. The HFEA may now additionally issue a licence for increasing knowledge about the development of embryos, increasing knowledge about serious disease, and enabling such knowledge to be applied in developing treatments. Details of research licence applications appear on the HFEA's website.

The **United Kingdom Xenotransplantation Interim Regulatory Authority** (UKXIRA), set up in 1997, also comes under the Department of Health (www.advisorybodies.doh.gov.uk/ukxira). It advises on safety, efficacy, animal welfare (in liaison with the Home Office) and any other preconditions for xenotransplantation into humans (the proposed transplantation of animal tissue and organs to treat human disease), the acceptability of specific applications, and strategic research in the area. Like the insurance committee, it was set up at a time of high interest in the technology, and found rather little to do, for reasons we outlined in chapter nine. It was wound up in December 2006.

In terms of food and agriculture in the UK, the **Agriculture and Environment Biotechnology Commission** was wound up in 2005, but their website (www.aebc.gov.uk) still gives access to reports and information and consultation on GM crops. Responsibility for safety of GM foods falls to the **Food Standards Agency** (www.food.gov.uk/gmfoods), which takes into account the findings of the **Advisory Committee on Novel Foods and Processes** (www.acnfp.gov.uk).

In the US

The **US National Institutes of Health**, part of the Department of Health and Human Services, operate an **Office of Biotechnology Activities**, which takes in a cluster of committees, currently including the regulation of recombinant DNA and gene transfer, the National Science Advisory Board for Biosecurity, and the **Secretary's Advisory Committee on Genetics, Health and Society** (www4.od.nih.gov/oba).

Of these, the Committee on Genetics, Health and Society is probably the most important for our interests, and has reported recently on genetic discrimination, direct-to-consumer marketing of genetic tests, and genetic education for health workers. The NIH's **Recombinant DNA Advisory Committee**, set up back in 1974 to oversee the hot new technology of gene splicing in the lab, still exists, but is concerned solely with technical details of hazard guidelines (www4.od.nih.gov/oba/rac/aboutrdagt).

One defunct committee is worth noting as an example of the political tangles around some areas of research. The NIH's **Human Embryo Research Panel** was set up in January 1994. The group classified human embryo research into three categories: acceptable; requiring additional review; and unacceptable. It also drafted guidelines for the review and conduct of acceptable research. The Advisory Committee to the Director of NIH unanimously approved the report, but President **Bill Clinton** blocked it. His verdict: "I do not believe that federal funds should be used to support the creation of human embryos for research purposes, and I have directed that NIH not allocate any resources for such research."

In the US, the **Food and Drugs Administration** is concerned with xenotransplantation, and operates a **Xenotransplantation Action Plan** to develop effective regulation. The Department of Health and Human Services (DHHS) set up the **Secretary's Advisory Committee on Xenotransplantation** (SACX) in 1999. SACX was wound up in 2005. A bit earlier than its British counterpart.

More broadly, of course, the US has a federal system, and individual states may take their own initiatives on particular technologies, or frame their own legislation. If you should need to know about these, the National Conference of State Legislatures operates a **Genetics Legislation Database**, which contains information on genetics bills and related research issues such as **stem-cell research** (www.ncsl.org). The database includes bills considered in state legislatures from 2004 to 2006. NCSL says it updates the database at least once a month.

Medical products eventually fall under the jurisdiction of the **US Food and Drugs Administration** (www.fda.gov), a massive bureaucratic machine. Genetic tests, for example, fall under the FDA's Centre for Devices and Radiological Health, and its Office of In Vitro Diagnostic Device Evaluation and Safety.

In Europe

The **European Medicines Agency** regulates clinical trials in the European Union, and handles gene therapy via its **Emerging Therapies and Technologies Division** (www.emea.eu.int/htms/human/itf/itfintro). The **European Food Safety Authority** (EFSA) has the general remit that its name suggests, and provides scientific advice to European Union Institutions and Member States on the safety and risks for human and animal health and the environment of GMOs. EFSA publishes its advice on its website: www.efsa.europa.eu/en

The European Commission takes more general advice from the **European Group on Ethics in Science and New Technologies,** an independent group with a wide remit. Their website (ec.europa.eu/european_group_ethics/index_en) is a good starting point for links to other EC activities in biotechnology, genetic manipulation, and its regulation. Their own recent reports include opinions on patenting stem cells, genetic testing in the workplace and umbilical-cord blood banking. Each opinion is preceded by a round table discussion involving representatives of the EU, institutions of the European Union, experts, NGOs, patients and consumer organisations and industry.

Help through the jungle

Industry is rather better served than the public with systematic information, and two of the most useful sites are aimed at biotech companies trying to find their way through the maze of regulations they have to thread on their way to market.

The **UK biotechnology regulatory atlas** is a useful one-stop shop for information in this area, and covers key US and European agencies as well. It is maintained by the **Laboratory of the Government Chemist** on behalf of the Department of Trade and Industry. At the time of writing it was on a very sluggish web server, but if you are patient the information is probably there. Make that coffee and go to plus.i-bio.gov.uk/ibioatlas. **The Genomics Gateway** at the University of Bradford (www.brad.ac.uk/ acad/sbtwc/gateway) rounds up international regulation of biotechnology – including a section on social and ethical issues which reproduces UN declarations on genetics and human rights and suchlike.

In the US, the website of the **United States Regulatory Agencies Unified Biotechnology** works very smoothly. It gives the lowdown on the roles of the US Department of Agriculture, Environmental Protection Agency and Food and Drug Administration.

Beyond governments

Apart from the Nuffield Council, all the bodies just listed are attached to government. That gives them a direct influence, of course, but there are plenty of organizations outside government which also try to have their say about the future of genetic technologies. General consumer, health and environmental organizations frequently keep a watchful eye on relevant aspects of genetics. Among those dedicated to genetic issues alone, it is worth checking out the following.

In the UK

GenewatchUK (www.genewatch.org) describes itself as a not-for-profit group that monitors developments in genetic technologies from a public interest, environmental protection and animal welfare perspective. The group issues regular reports and briefings, and lobbies MPs and ministers.

Its campaigns at the time of writing included encouraging people to "reclaim their DNA" from the police if they are on the DNA database but have not been charged or convicted of a serious offence, and calling for legislation to prevent employers using genetic test results to decide who gets a job or pension.

Comment on Reproductive Ethics (www.corethics.org), founded in 1994 is a pro-life group which lobbies on new reproductive technologies. It is mostly against them: its main tenet is "absolute respect for the human embryo". **Human Genetics Alert** (www.hgalert.org) is a small watchdog group, based in London, funded chiefly by the Joseph Rowntree Charitable Trust. It describes its stance thus: "we are not opposed to genetic research. However, we are opposed to some developments, such as genetic discrimination, cloning and inheritable ("germ-line") genetic engineering of human beings".

The **Genetic Interest Group** (www.gig.org.uk) is an umbrella organization for patient groups representing those with genetic disorders. It has 130 member charities, and works to raise awareness and influence policy and research. Its main goal is to expedite treatment for people who often have severe or fatal conditions for which there is currently no remedy. As well as general information, and briefing on current campaigns, it is also an excellent gateway to information about genetic disorders and care groups.

In the US

The **Coalition for Genetic Fairness** was founded in 2000 to campaign for legislation barring the misuse of genetic information in insurance and employment. The first members were civil rights and patient and health care organizations, but it now includes industry groups and employers. At the time of writing, the latest version of a **Genetic Information Nondiscrimination Act** (GINA) had been introduced and passed unanimously by the Senate. An identical bill, HR.1227 has been sent to committee in the House and has more than 200 co-sponsors. Read the latest at www.geneticfairness.org/act

The **Council for Responsible Genetics** (www.gene-watch.org – not to be confused with the UK group) is a Boston-based group founded in the 1980s which campaigns on a wider range of issues but has had a long-standing interest in human genetics. It aims to foster public debate about the social, ethical and environmental implications of genetic technologies, saying "CRG works through the media and concerned citizens to distribute accurate information and represent the public interest on emerging issues in biotechnology". Its history is of leaning more to the left than to environmentalism, as suggested by its "core principles":

▶ The public must have access to clear and understandable information on technological innovations.

▶ The public must be able to participate in public and private decision making concerning technological developments and their implementation.

▶ New technologies must meet social needs. Problems rooted in poverty, racism, and other forms of inequality cannot be remedied by technology alone.

It has a number of academic advisers who have long been involved in biomedical science and public policy. The group publishes a bi-monthly magazine, also confusingly known as *Genewatch,* and regular web briefings and policy statements.

The **Genetic Engineering Action Network** (www.geaction.org), despite its name, campaigns specifically about (well, against) use of genetic modification in agriculture. It is an umbrella organization for groups active on the issue, and supports local activists in many parts of the USA. If you want to find a link to, say, GMO-free Alameda, this is the site to visit.

International

GRAIN (www.grain.org) is an international non-governmental organi-zation (NGO), based in Spain, which promotes the sustainable manage-ment and use of agricultural biodiversity based on people's control over genetic resources and local knowledge.

These are just starting points. There are numerous other groups with more specific objectives. Some promote or campaign in specific areas like GM crops, or even particular kinds of crops (www.banterminator. org). Others are concerned with disability, with specific genetic diseases (from Huntington's to sickle cell and cystic fibrosis), with use of genetic information, and so on.

Getting it right?

One of the most comprehensive examples of new-style consultation was a nationwide initiative in Canada in 2001 to sound out views on xenotransplantation. The Canadian Health Ministry laid out three quarters of a million dollars (about US$500,000) to get an answer to the question: "Should Canada proceed with xenotransplantation, and if so, under what conditions?"

The consultation, which was carried out independently by the Canadian Public Health Association, included just about every method they could think of. There was a national telephone survey, a questionnaire posted on the web, a survey of "stakeholders", invitations to write letters and six "citizens' juries".

The citizens' juries were held in six regions (Canada is a big country). Each involved 15–25 people, recruited after letters of invitation were sent randomly to 2500 in the chosen region. Some of those expressing an interest were then selected for rough balance in age, sex and education. Each group then met over two days, first of all in a public session, during which they and a wider audience interrogated experts on the topic, then for closed sessions during which they considered their own views.

This is just one example of how to go about getting what has been called "the social equivalent of informed consent". Or perhaps "informed denial", as the answer to the question in this case turned out to be "no". You can read about the whole thing at www.xeno.cpha.ca/english/about/page1

How can you get involved?

Ordinary citizens ought to have the final say about which technologies are developed, and how they are used. It is our future which will be affected, after all. When the technologies are as personal and, at least potentially, as powerful, as those likely to grow from genetic science, the demand for democratic participation is stronger.

So the ideal system would be one with plenty of opportunities for debate – with scientists on tap, but not on top – one which fed into the more formal side of regulation and legislation in ways which everyone agrees make a difference. As people always disagree about the actual decisions, so they will also differ about whether this ideal is ever achievable. But there have been plenty of recent experiments with forms of discussion that involve ordinary members of the public in debating new technologies, and many of them have been organized around issues in genetics.

These experiments take place alongside more traditional forms of public involvement, such as responding to consultation documents from committees or government departments, monitoring the work of expert committees – much easier nowadays as they publish more about themselves – or submitting evidence to parliamentary or congressional committees.

The newer initiatives take a number of forms, going by the names of **consensus conferences**, citizens' juries, deliberative polls, or just **public consultation** or "dialogue". Some are one-offs, some part of continuing efforts. The **European Commission**, for example, has a website which acts as an entry point into fairly regular consultations about the "biosociety" the EU sees as a key part of future economies: ec.europa.eu/research /biosociety/forum/bioforum_en

It is worth emphasizing that most of this consultation and debate remains focused on what we do with new technologies *once they have come close to application*. There are far fewer opportunities for the general public to get involved in discussion of what kinds of technologies ought to be developed at an earlier stage, or what kinds of science might be funded. But it is always worth investigating what **research funding agencies** are up to, because the labs they support – along with their private-sector equivalents – are where the next generation of technological possibilities will be created.

The **International Center for Scientific Research** (www.cirs-tm.org) has compiled weblinks for research organisations, institutes and universities in 122 countries. They are searchable by country and by subject. So why not log on to your local medical or agricultural research agency and see if they are interested in public input about research priorities. If not, ask them why not.

How do we know what people really think?

It is easy to agree that decisions about potent new gene technologies should take public views into account; harder to know what people actually think. Opinion polls and surveys help. Ask lots of people the same question and you can get statistically useful results. The **UK Human Genetics Commission**, for example, has to mull over the fact that, according to a survey it commissioned, nine out of ten people agree that genetic developments could and should be used to diagnose and cure disease; at the same time, a third of the sample of over 1000 British people agreed that research on human genetics was "tampering with nature", and was unethical. Plenty of other polls yield similar findings, which hint at a deep ambivalence about the whole enterprise.

Broad-brush results like this do not seem a particularly helpful way of summarising what "the public" thinks. Stop people on the street and ask them a pre-set list of questions, and they will probably try and answer them, even if what they are really thinking is that they have no interest in the topic, and have never considered it before. More detailed, surveys – like the "**Eurobarometer**" exercises regularly carried out in all the nations of the European Community – suggest people think differently about medical and agricultural applications of genetics, for example. "**Red**" biotechnology, applied to curing diseases, gets consistently higher approval scores than "**green**" biotech, applied to crop plants or animals. (This rather neat distinction has now been complicated by the introduction of a third category they call "**white**", for industrial biotechnology – which is also broadly speaking finding support from European publics). The Eurobarometer series began in 1991, and the survey is repeated

GM Nation

Elaborate consultations can inform a debate, but will not defuse controversy. A nationwide debate was organized in Britain about GM foods in 2003. This move from the UK government came after intense public disputes about whether GM crops or their derivatives should be allowed at all, whether products containing them should be labelled as such, and so on (see chapter 10).

Alongside a review of the science behind GM, and another which looked at the costs and benefits, the government instituted an extensive, though quite short, debate on the whole subject. Among other things, there were some six hundred meetings across the country, in village halls and conference centres, as well as in-depth discussion groups. Lots of citizens looked at special briefing materials and a website, and more than 37,000 people registered their views.

As the final report put it, people on the whole "expressed strongly the belief that GM technology and GM food carried potential risks and a majority rejected any suggested benefits from GM, except to the companies which promoted it. Such attitudes varied in intensity but they did represent the majority in all sections of the active participants in the debate."

Findings like this feed back into further discussions. In this case, those in favour of GM crop developments protested that the debate had been too short for people to learn about them properly, and that the meetings were poorly publicized and packed by organised groups of opponents. Critics of GM responded that there were plenty of participants, with genuinely concerned opinions, and that having organized the debate the government should listen to the results.

The government, through the Department of Environment, Food and Rural Affairs, reaffirmed its commitment to consider proposals to plant GM crops on a case-by-case basis. You can read about the debate at www.gmnation.org.uk and study the response at www.defra.gov.uk/Environment/gm/crops/debate/pdf/gmdialogue-response.pdf

every couple of years. You can find the latest results at www.ec.europa. eu/research

As far as medical applications of genetics go, attitudes differ around the world according to the nature of the practice: whether it's **genetic screening** (often favourably rated, but viewed with more suspicion when insurance companies are mentioned as well), **genetic diagnosis** to help more precisely targeted drug prescription, or real **genetic manipulation**. And there tends to be a kind of hierarchy of organisms. When surveys list different applications of genetic manipulation, then altering bacteria prompts less suspicion than efforts to modify animals and humans. Plants are a more controversial subject in Europe than in the US. Latest figures suggest that American survey respondents were most "comfortable" with the modification of plants (a 5.94 average rating), followed by microbes (4.14), animals used for food (3.73), insects (3.56), and animals used for other purposes (2.29). They were least comfortable with the genetic modification of humans (1.35). When asked specifically about genetic modification of animals, more than half (57 percent) of the Americans surveyed opposed it; only one third (32 percent) was in favour. Even so, human genetic manipulation to treat or prevent disease is (prospectively) acceptable to quite a few people.

Other developments tend to draw a stronger reaction. Polls in many countries show a consistent opposition to reproductive cloning – an issue where it seems relatively easy for legislators to stay in tune with public opinion (many countries, including the UK, now have laws that make any attempt to clone humans a punishable offence). **Stem-cell research**, of the kind that involves embryonic material, draws fire in countries where anti-abortion sentiment is strong. However, immediate reactions to a new issue like this can change relatively quickly. The latest (2006) summary of public understanding and attitudes from the **US National Science Foundation** suggested that opposition there to medical research that uses stem cells from human embryos has declined. In 2004, 36 percent of those surveyed said they were opposed to this type of research, down from 51 percent in 2002.

Genetic technologies take many forms, but the one which is likely to produce the hardest problems, and which draws the strongest reactions, is altering the genetic make-up of human beings. Still more science fiction than science fact, it has nevertheless featured in a number of studies trying to probe whether people want to live in a brave new world of genetic control. The most determined efforts have focussed on near-term medical uses – in other words **gene therapy**. We'll round off this section by looking at some of these in more detail.

A national UK survey on gene therapy 1999, carried out on behalf of the **Wellcome Trust**, is helpful because it asked so many detailed questions. Although quite old now, most of the findings have held up over the years when they have been checked against the results of later surveys. This study found that awareness and knowledge of gene therapy were both fairly low. The survey results are based on face-to-face interviews with a representative sample of just under 700 adults across the UK in late 1999. Most of them had heard of genetically modified foods and cloning, but awareness of gene therapy was quite a bit lower. Asked if they had heard about "curing diseases through genetics" within the last month, only 41 percent said yes. This figure was down to 31 percent among readers of popular newspapers, but rose to 64 percent among readers of more upmarket titles.

Almost four fifths of these people reported that they received little or no information about genetics, and around half the sample had not discussed it with anyone else in the past year. A simple genetic knowledge quiz suggested that most people had a little basic knowledge of genetic facts, but more often resorted to "don't know" when asked about the details.

Not surprisingly, knowledge of the current state of gene therapy was also poor. More than half the sample believed that some diseases could already be treated by altering genes. There is not much evidence that this is true now and there was certainly none in 1999. Some scientists' and journalists' wishful thinking is evidently catching.

PERCENTAGE OF PEOPLE AGREEING WITH GENETIC RESEARCH

Positive:

Genetic treatments for illness will do a lot to reduce human suffering	76
In the end, research into human genes will do more to help us than to harm us	68
People worry too much about the risks of changing human genes	43

Negative:

It is better to try to cure illness without changing people's genes	74
Changing a person's genes is too risky, whatever the benefits might be	49
We should never interfere with people's genes	31
It would be better if we did not know how to change people's genes at all	28
Changing genes should be forbidden as it is tampering with nature	26
Scientists should not look for genetic cures, because the world will become too overpopulated	20

Source: national UK survey on gene therapy, Wellcome Trust, 1999

They may not have known much about it, but the people in this survey did have opinions about genetic research and gene therapy. A big majority thought that genetic treatments could reduce human suffering, and that research would do more good than harm. However, an equally large majority also agreed that it would be better to cure illness in other ways. And quite a few felt that we should never interfere with people's genes, and it would be better not to find out how to do it. With these general attitudes, responses to questions about a range of specific scenarios for genetic alteration were interestingly varied.

TRUE OR FALSE?

Identical twins have the same genes	**65**	17	18
Whether a couple have a boy or a girl depends on the woman's genes	15	**64**	21
Half your genes come from your mother and half from your father	**61**	19	20
Down's Syndrome is an inherited disease	22	**58**	20
Children look like their parents because they have the same type of red blood cells	15	**58**	26
Most cells in our body contain a copy of all our genes	**57**	7	36
There are test tube babies who grew entirely outside the mother's body	31	**47**	22
Genes of all living things on earth are made up of different combinations of only 4 or 5 chemical building blocks	**28**	12	60
We have around 150,000 different chromosomes which contain our genes	31	**13**	56

Note: numbers listed are percentages; the three columns on the right correspond, respectively, to the answers "Yes", "No" and "Don't know"; the correct answer is indicated in bold

Source: national UK survey on gene therapy, Wellcome Trust, 1999

If the target condition were a serious illness, and only body cells were affected (**somatic gene therapy**), a large majority would consider genetic alteration acceptable – 92 percent for cystic fibrosis, and 82 percent for heart disease. A two-thirds majority would even allow changing genes to remedy baldness. Around the same proportion would agree to gene treatments for those merely at risk of heart disease, who had no symptoms. But moving into the realms of enhancement – improving memory – only got the thumbs up from 34 percent. If the change affected reproductive cells (**germ-line therapy**), so that the person's *children* would be spared disease, have a full head of hair or better memories, approval fell. But three quarters of the sample still welcomed this prospect in the case of cystic fibrosis. Altering genes in the womb was always the least acceptable option. These are mostly answers to hypothetical questions, and so they may not predict how people will feel if any of these procedures actually were to happen. But this picture of current attitudes to gene therapy is pretty consistent with other more recent surveys. All show strong support for the idea of gene alteration – even germ-line alteration – to prevent or cure disease. At the same time, there is general wariness about the possibility of gene alteration for non-medical use.

For example, the regular **British Social Attitudes** survey, carried out in 2003, found that 81 percent of respondents said that genetic alteration should "definitely or probably be allowed" to reduce someone's chances of getting breast cancer. But more than half the people asked would also allow it in order to make someone less aggressive or violent. Worryingly, nearly one person in five (18 percent) said they would be willing to see it used to make someone heterosexual rather than homosexual – slightly more than would countenance choosing the sex of an unborn baby (15 percent).

Other surveys from overseas record similar findings. A nationwide telephone survey in the USA in 2002 for the **Genetics and Public Policy Centre** at Johns Hopkins University found the same split between opinions about genetic engineering to avoid disease and to create "desirable" traits. Disease avoidance prompted 59 percent approval and 34 percent disapproval. Use of genetic techniques to promote strength, intelligence or attractiveness, on the other hand, was condemned by 76 percent of this sample, and approved by only 20 percent.

This study also found that knowledge of gene therapy was pretty sketchy in the US, as in the UK, with around a quarter of people believing that it was already possible to prevent a genetic disease using genetic engineering. And, as in the UK, approval for efforts to deal with disease extended to germ-line alteration. This time, the 59 percent approval figure was in response to the following question: would you approve or disapprove if parents were offered a way to change their *own* genes in order to prevent their children from having a genetic disease?

In Australia, a nationwide survey of attitudes to biotechnology conducted in 2003 – the latest in a series – found that 71 percent believed that gene therapy would "improve our way of life". In fact a similar proportion (21 percent) of people to the US believed that gene therapy was already doing so. Another 24 percent thought it would bring improvements within five years, and 28 percent in 10 years. If that seems like approval, then bear in mind that 33 percent of those asked Down Under were so convinced gene therapy exists that they said they thought it was already making things worse. A reminder that when it comes to opinions rather than knowledge people like to oblige pollsters by answering the question.

Genes and culture

Getting a handle on genes

David Cronenberg's 1986 remake of the classic 1958 horror film *The Fly* had famously horrid special effects. It also updated the original premise – an accident with a matter transmitter causing the scientist Brundle to transmute into a fly – with a lot of modern gene talk. When Cronenberg's character asks his computer what is happening to him, it succinctly informs him that there has been "integration at the molecular genetic level". It's one moment in one movie, but it's the tip of the genetic iceberg in popular (and not so popular) culture.

A 2003 Harris poll found that 60 percent of adults in the United States selected the correct answer when asked "what is DNA?" (the genetic code for living cells), and two thirds chose the right answer when asked "what does DNA stand for?" But knowing about DNA is not the same as knowing how it works, or how important it is – what it all *means*. In this chapter, we take a look at some of the ideas and images about inheritance and genes in circulation, from scientific papers to comic books.

Thoughts about genes

Genetics is a science, and the formal language of science is sober, restrained and precise. But scientists also speak the same language as everyone else – and they use it about their own subject. **Watson and Crick**'s landmark paper in *Nature* about the double helix was titled "A Structure for Deoxyribose Nucleic Acid". They proclaimed to lunchtime drinkers in a Cambridge pub that they had found the secret of life.

And when scientists get excited about the discoveries they have just made, or might make one day soon, journalists get excited too. Together, they seem to have built up a picture in which genes do not just control the order of amino acids in proteins. They seem to dictate every last characteristic of an organism. In our favourite creatures, humans, they are are held responsible for all kinds of behaviour, as well as physical or physiological traits. As sociologist **Dorothy Nelkin** and historian **Susan Lindee** noted in their book *The DNA Mystique* in 1995, if we are to believe the news media, "there are selfish genes, pleasure-seeking genes, violence genes, celebrity genes, gay genes, couch-potato genes, depression genes, genes for genius, genes for saving and even genes for sinning". The pair reckoned this picture of all-powerful genes was underpinned by predictions that an individual's tendency to healthiness or disease would be revealed by the Human Genome Project. It was summarized in one-liners like **James Watson**'s declaration that "our fate is in our genes". Thanks, Jim.

At the same time, more scientific – or at least more scientifically *derived* – ideas about genes and the extent to which they make you the person you are coexist with older notions about **inheritance**. And all of them are exuberantly reworked in fiction and drama. Genes are linked with topics of abiding human interest: sex, death and identity. So it is no surprise that genetics excites novelists, videogame makers, dramatists and filmmakers.

Filmmakers love revisiting powerful stories from the pre-genetic era, such as *Frankenstein* or *The Island Of Dr Moreau*, or fashioning new plots for our amusement, like the film *Twins*, which pairs **Arnold Schwarzenegger** and **Danny DeVito** as ill-matched siblings. And they help shape our ideas about what might happen. Somewhere in this entertaining mix of science, fiction, science fiction and folklore, possible futures are being forged.

James Watson

Blue blood and giraffes

We also still hear regular echoes of much earlier ideas. For most of history, the idea of the gene would have sounded outlandish. Invisibly tiny entities, distributed throughout our bodies, which are shuffled randomly before being passed down to the next generation in a single copy? Yeah, right! And some ideas from the past seem equally outlandish to us. **Aristotle** believed that the foetus was formed by the action of the semen on menstrual blood, the greater "innate heat" of the male principle impressing itself on the formless matter provided by the female. This is an idea rarely heard in the way we talk about **inheritance** now. Yet it influenced animal breeders for millennia. And it is still not hard to find people who believe that the father's characteristics are more important to the child than the mother's.

Another idea which is hard to shake off is that the child's characteristics are a blend of his or her mother and father. This is a distant descendent of a notion from another influential ancient thinker, **Hippocrates**, generally regarded as the founder of Western medicine. The Hippocratic notion of *pangenesis* held that there was a contribution from both male and female semen, which was drawn from all parts of the bodies of both parents by a mixing of the humours, warmed nicely during sex. The contribution from each parent could vary in different organs, so some might resemble the father more, some the mother. **Charles Darwin** came up with a theory pretty similar to this (involving minute "**gemmules**" in the tissues) when he was struggling to understand what theory of heredity could underpin his ideas about natural selection. And when we read today in the *Weekly World News* headlines such as "Farmer breeds cow with pig" to raise a hybrid with meat that tastes like a bacon burger, it is the idea of **blending inheritance** which makes the story sound faintly plausible, as all good supermarket tabloid stories should. Hollywood still likes the idea, too. Seth Brundle's transformation in the David Cronenberg version of *The Fly* is the concept of blending inheritance returning with a vengeance.

Some earlier takes on what we now define as inheritance leave less trace on how we talk about it now. The long popular theory of "maternal impressions" is typified by the story in Genesis 30 of how a youthful Jacob gets his own back on Laban, who has reneged on a deal to give him all his striped or spotted cattle by removing them from the flock in Jacob's absence. Jacob takes various rods from the trees of the forest, whittles them until the white sap inside them is visible and piles them in front of the cattle shortly before they are due to conceive. Lo and behold, the cattle are born "ringstraked, speckled, and spotted". In other words, what the cattle saw gave their offspring their markings.

In medieval times, many swore that almost anything which swam into the mother's consciousness – a picture on the wall glimpsed during lovemaking, an animal which sat on her lap – could affect the child. Few believe this nowadays, of course. But if you were to eavesdrop on almost any family gathered around the cot you might hear comments along the lines of "Look, he's got his father's nose". They hark back to another old idea – **pre-formationism**. It sounds as if we believe in the transmission of an actual miniature nose.

More recent scientific ideas which have been officially discredited also cling on in popular beliefs, their intuitive appeal outweighing scientists' insistence that all the evidence is against them. The great French naturalist **Jean Baptiste Lamarck** advocated inheritance of acquired characteristics as the mechanism of evolution in the first half of the nineteenth century. As well as giving a neat explanation for how creatures end up adapted to their environments – the length of giraffes' necks is still the favourite classroom example – this accounted for plenty of everyday observations. Look, the village blacksmith has great, bulging forearms, just like his father, the blacksmith before him!

But after Charles Darwin's evolutionary theory met Mendel's genetics, inheritance of acquired characteristics was out, scientifically speaking. The **germ plasm**, later refined to refer to the genes in the chromosomes, was found to be immune to influence from the environment after conception (although the discussion of epigenetics in chapter 4 provides some very recent ideas about exceptions to this).

And so we learn in school. Ask schoolchildren whether athletes will have super-fit children, or if a puppy which loses a leg in an encounter with a car will have crippled offspring, and the ones who have been listening to teacher will say (correctly) that the answer is no.

French naturalist Jean Baptiste Lamarck

Yet even if we are well-versed in the modern, scientific story about inheritance, it coexists with lots of older ideas whose influence is hard to shake off. And fluency in the language of genetics does not mean a person will not speak of some trait being "in the blood". It is a good example of how cultural inheritance is harder to track than genetic inheritance.

Good and bad metaphors

How genes, and their workings, have been perceived has changed a few times over the last hundred years. To some extent it has moved in step, or maybe a bit behind, the changes in scientific ideas about genes we discussed in chapter four. Media analyst and historian **Ruth Condit**, in her book *The Meanings Of The Gene* finds that ways of depicting genes in public prints have gone through four phases. To begin with, the genes were depicted as part of the **germ plasm**, the crucial stuff that was passed on to the next generation, and how they worked was seen in terms of stock breeding. The gene was a rather abstract kind of thing, and detailed mechanisms were beyond the scientists of the time. Furthermore, it wasn't regarded as important anyway. In 1926 the great **Thomas Morgan**, pioneer of gene mapping in the fruit fly, wrote a technical book called *The Theory Of The Gene*. It turned out that there wasn't one, really. "The theory of the gene is justified", he declared blithely, "without attempting to explain the nature of the causal processes that connect the gene and the characters". Don't worry how genes do stuff, in other words. They just do.

Early popular accounts of genes tended to follow Morgan's advice. In fact, they didn't try very hard to explain how we knew *where* genes are, or even that they exist at all. A generously proportioned tour of biology published in England in 1938, **Odham**'s *The Miracle Of Life*, just says: "... the gist of it is something as follows: chromosomes are elastic-like rods in the

From factory to theatre

Authors who write whole books about genetic topics usually get to develop their metaphors in the most painstaking detail, often by contrasting old and new. A nice example is Steven Pinker's attempt to portray genes as part of a constantly shifting scene as an organism develops, in his book *The Language Instinct*.

"Imagine that what the genes control is not a factory sending appliances into the world, but a machine shop in a thrifty theatre company to which props and sets and materials periodically return to be dismantled and reassembled for the next production. At any point, different contraptions can come out of the shop, depending on current need." Imagine indeed...

substance of the nucleus on which the genes are threaded rather as beads on a wire. The genes are bodies so small that they cannot be seen singly even under the most powerful miscroscope. We do not know whether the chromosomes consist wholly or partly of genes. But we do know that the genes are always present in a definite order". Hard, perhaps, to see beads on a wire, of some unknown substance, as all-powerful shapers of life.

By this time, popular writing was in the era Condit calls "family genetics". Newspaper accounts of the new science centred upon the hopes and fears of individuals, and the main metaphor was the gene as a **particle** – a kind of atomic theory of genetics. What it might be made of was not much discussed. After 1953, however, we knew there was a DNA molecule that contained information. The second half of the twentieth century was, of course, when the science of DNA was developed, and it was understood in terms of codes and language, translation and transcription. Some historians have seen this shift in terminology as a result of a cold-war preoccupation with cybernetics and systems analysis, command and control. But wherever it came from, it is now deeply embedded in the technical language of biology as well as in popular writing.

Finally, says Condit, there was another subtle shift from around the mid-1970s, when **medical genetics** hogged the headlines, the idea of the genome was in the forefront, and the dominant metaphor used became the "blueprint" – the plan on which the cell or the body is built.

Have a nice DNA

This quick sketch is only one brief history. It does make the point, however, that there has been a range of metaphors applied to genes and DNA. The most important way of thinking, though, and the one heard and read most often, is probably the idea of genes as **information**. This is not surprising in today's so-called "information society", although the metaphor goes back much further. Try, for example, this explanation of how the same genes could end up in cells which do lots of different things. "As a result of the delicately adjusted processes of cell division the whole complex [the genes] is distributed together, much as a number of copies of a newspaper are printed, all alike, and distributed. Some sets go to the hands, some to the feet, and they work according to their situations. It is like copies of one and the same popular newspaper going here into the hands of the sporting man, who reads merely the racing news, here to the speculator who concentrates on the City column; Mr Everyman runs his eye ineffectively over most of it and concentrates on the crime and the cricket, while Mrs

Everyman (when he has done) consumes the Court and Social intelligence and the fashion and household pages."

A "popular newspaper" offering "Court and Social news" gives us a clue that this is from the 1930s: **H.G. Wells**, his science writer son G.P. Wells and the biologist Julian Huxley came up with this image of DNA as text in the hugely successful *The Science Of Life*, years before the double helix was even dreamt of. Still, it was in the computer age that the ideas that information is the essence of everything, and DNA is the essence of an organism, came into their own. **Richard Dawkins**, as usual, puts it across with a literary flourish. "It is raining DNA outside ... It is raining instructions out there; it's raining programs; it's raining tree-growing, fluff-spreading algorithms. That is not a metaphor, it is the plain truth. It couldn't be any plainer if it were raining floppy discs".

Those floppy discs date the image a bit – this is from *The Blind Watchmaker* in 1986. Dawkins had put it even more forcefully ten years earlier in his first and most famous book, *The Selfish Gene*: "We are survival machines – robot vehicles blindly programmed to preserve the selfish molecules known as genes".

There are a few problems with all this if a non-geneticist wants to get an idea about how genes *actually work*. These metaphors mostly just convey the idea that genes are a power within, which suffices for many news stories, and for many science fiction writers. Popular science writers try to go one better. It is easy to figure that genes by themselves do not actually do anything. An isolated gene is, indeed, as useless as a computer program with no computer. You grasp some idea of what goes on in the cell by extending the metaphors a bit. So the molecular machines which use the DNA have often been depicted as little bits of information technology. Complicated assemblies of enzymes, "read" the DNA, transcribe it into RNA, and then take the RNA read-out from a gene and translates it into a custom-made protein.

This makes it sound as if the cell is packed with clever technology (which in a way it is), and that it is rather like our technology (which is a bit more of a stretch). The efforts of biotechnologists can then be depicted as a more or less logical extension of nature's own technology. An impressively elaborate example of this occurs in biologist **Robert Pollack**'s popular book *Signs Of Life: The Language And Meanings Of DNA*. As the title suggests, the whole book asks how far you can take the notion of DNA as language. When he explains recombinant DNA technology – as we try to do in chapter three – Pollack says it is really just like word processing. He describes how all the usual functions of copy, cut and paste can be carried out by the right combination of bacterial enzymes. Very neat. But it does suggest that delving into another organism's DNA and doing a spot of rewriting is a perfectly natural thing to do, rather than a rather startling departure from previously existing human-built technologies.

This vocabulary of cutting and pasting makes genes and the manipulation of genes sound rather straightforward. But a larger problem is that the whole metaphor, as metaphors tend to, starts to break down when you look more closely at actual genes – made of minuscule bits and pieces of molecules, and not at all like printed text. The scientific notion of the gene is still changing as knowledge advances, and it is harder and harder to define what a gene is. Time, perhaps, for some new metaphors for the rest of us to try on for size.

A nice suggestion from **Hans-Jörg Rheinberger**, Director of the Max Planck Institute for the History of Science in Berlin, is that the "whole battery of mechanisms and entities [constitute] what could be called *hereditary respiration*, or *breathing*" (our italics). This is charmingly poetic, but seems not really concrete enough as an aid to understanding. Nor does the suggestion that genes are "things the organism can do with its genome" leave you feeling much better informed.

Another way to sum up the complexity of what is going on inside cells is to suggest that, in the words of Australian philosopher of biology **Paul Griffiths**: "In the post-genomic era, agency is being relocated to the genome as a whole and perhaps beyond." What he means, we suspect, is that it is not genes which do stuff, but whole systems of genes and all their associated bits and pieces. This is a nice way of reducing the dizzying details of all the tagging, splicing, signalling and switching going on, but still seems too philosophical to achieve wide appeal.

What else might we try? Well, there are only so many things a living system might be like. The ones which have been offered so far are musical, ecological, geographic and social metaphors. The musical variations may not get very far, as they are still rooted in a static concept of information. The DNA becomes a musical score instead of a linguistic text. Someone, or something, might interpret it or use it for orchestration. But it is hard to place the conductor, as it were.

John Avise from the University of California, Davis, takes another tack. See the genome as a commune, he suggests: a tightly bound organization with an intricate division of labour. Or maybe it is a kind of ecosystem in which different genes fill different niches. Other tropes that have been used include **systems biology**'s focus on networks. A road network, for instance, is interconnected in ways which offer many different routes from A to B.

An individual gene might then be like a single road. Block it, and the traffic may still get through, though by a more circuitous path.

Probably the most striking image ventured so far comes from philosopher **Lenny Moss**. He considers two stages of gene expression which are regulated by impressively large assemblies of molecules: which in fact work closely together. First is transcription, in which the precise selection of exons for an **RNA transcript** depends on the presence (or absence) of a variety of transcription factors, which interact both with DNA and with RNA polymerase. The second is post-transcriptional splicing of the messenger RNA, which is the job of a remarkable agglomeration with the horrid name of the **spliceosome**. It has been unpacked to lay out an intricate array of five small nuclear RNAs and as many as 150 separate proteins.

The sensitivity and microscopic modulation of these operations is hard to grasp. Moss's best shot is his suggestion that the "decisions" about the final configuration of the mRNA are taken on the path from DNA to protein by "*ad hoc* committees". Think of each committee, he suggests, as a constituent assembly. The more members it has, the more information they can all draw on about recent history of events in the cell, and of interactions with other cells. The committee is a way of pooling experience before a kind of consensus is reached about what to do next.

These cellular assemblies offer an image of a dynamic and flexible operation which is – like at least *some* human committees – wiser than any of the individual parts. Are the things going on in cells really like this? Who knows? But it is a nice way to think about them – a long way from the simplicity, and maybe rigidity, of the central dogma. And it captures the idea that, if genes are us, they only make us on terms dictated by history and circumstance.

All these metaphors and analogies may seem a little awkward, and when they are extended they often are. The aspects of genes that they try to capture are important in popular science books and documentaries. But much of the rest of popular culture gleefully ignores them in any case. Lack of time, tight space, or just the urge to make a simple image or a powerful story tend to overrride any lingering doubts about whether genes do what the authors say they do.

Genes at large

This heading is not intended to conjure up images of genes rampaging across the countryside – though that does seem to be a common nightmare for some folk. But the G-word is invoked in lots of places outside of science articles. It crops up almost every time character and temperament are discussed, for instance. In a pre-World Cup football article in a newspaper, the Ukrainian team boasted of having the teachings of their former trainer "in their DNA". Jon Bon Jovi once explained that the enduring success of his group is that "self-belief is in the band's DNA".

Comics, cartoons, stories, plays, films and commercials make free with gene imagery. Sometimes the advertising pitch is again just about character or quality. BMW cars were sold on the basis of a "genetic advantage". Bijan fragrances of Beverley Hills launched their "DNA" perfume in 1993 (though it doesn't have any in it). Other cosmetics do, or at least they claim to. There are a host of face and skin creams that have added DNA to the list of vaguely scientific words used to persuade willing purchasers that smearing stuff on their epidermis will maintain their youthful good looks. More interestingly, especially in the last fifteen or twenty years, visual artists have picked up on the idea that there is something intriguing about the relation between genes and bodies. This kind of cultural production, high and low, is inexhaustible, and doesn't lend itself to an easy summary. Below are brief notes on some novels and films worth checking out. Here, we mention just a few key themes.

Chemist **Carl Djerassi** has written a series of novels and, more recently, plays, intended to impart information about science and scientific practice. He calls them "scientifiction" – you will learn details of the **polymerase chain reaction**, for example, if you happen to read *The Bourbaki Gambit*, probably the best of a mixed bunch. More often, fictions explore consequences of particular possibilities such as artificial reproduction or genetic alteration. Cloning has attracted many writers because – as we described in chapter 11 – it taps into old stories about *doppelgängers* and impersonation, as well as offering plots which can explore sexual rivalry, identity, parent-child relations and other themes which are the stuff of drama.

Other art forms, it could be argued, have not yet reached this level of sophistication about genetics. Take visual arts, for instance. DNA found its way into **Salvador Dali**'s surreal landscapes in the late 1950s, but the real growth of DNA-inspired art has been in the last couple of decades. On examination, however, some of the more celebrated works seem less profound than one would hope for. The British artist **Mark Quinn**, for instance, produced a portrait of genome scientist **John Sulston**. It was not a conventional image, but a collection of bacterial colonies containing DNA spliced from a portion of Sulston's own genome. According to the National Portrait Gallery, who exhibited the work, "although abstract in appearance, in reality the portrait represents the most exact likeness possible of the sitter." This seems a rather over-simple representation of the notion that DNA is us, and once you have said that, there's not much more to get out of the portrait.

Similarly disappointing is **Eduardo Kac**'s widely discussed fluorescent bunny, supposedly produced by incorporating the gene which codes for phosphorescent protein found in a certain jellyfish into rabbit embryos. His website (www.ekac.org) features images of the green, glowing bunny, together with an extended discussion of his plan to do the same for a dog. Indeed, he announces that he is practising a new art form, **transgenic art**, which will create organisms that can be taken home by the public to be grown in the backyard, or raised as companions.

This is interesting, up to a point, but less so when you discover that it is all pretty much the same as the project carried out by scientists at the Oregon Regional Primate Research Center in 2001 to create **ANDi**, the transgenic monkey. ANDi ("inserted DNA" backwards, you see) was the first genetically modified primate, and having also been implanted with the jellyfish gene, which is used because it is easy to identify – though he doesn't actually glow green. (He also graces the inside cover of this book.) So while Kac's bunny has provoked some comment, it is a little hard to see what he has added to the debate triggered by ANDi's birth by calling his project "art" instead of "science".

There are things you can do with DNA at the more commercial end of art, too. Go to 11DNA.com and for just $390 you can have a DNA finger-print prettied up and framed to make an artwork which is really "you". The price includes a DIY saliva-collection kit, and you get to choose the size and colour of the finished piece. Then again DNA helix bracelets from theDNAstore.com cost a lot less, look nicer, and you can wear them. They may also be less likely to make friends doubt your sanity.

These are just a few out of a welter of DNA-related art projects of recent years. (You can read about many more in Suzanne Anker and Dorothy Nelkin's book *The Molecular Gaze: Art In The Genetic Age*.) It seems fair to say, though, that few gallery artists have yet come near the impact of gene-related popular works like *X-Men*. The narrative media still seem to be the most promising source for stimulating stuff which can be used to provoke thoughts about genetic futures, real or imagined.

PART 7
RESOURCES

In print and online

Fiction

Genetic engineering, mutants, post-human evolution and cloning are staples of science fiction that proliferated after World War II. Cloning also has its appeals for thriller writers and literary novelists. This chronological list is restricted to books still in print. (One of us has fond memories of Russell Braddon's *The Year Of The Angry Rabbit*, in which a mutated virus released in Australia does very bad things indeed, but we doubt you'd be able to find it now.) Below are the landmark novels, plus some others worth tracking down.

Frankenstein Mary Shelley (1818) Not strictly about genetics, but the youthful Mary Shelley's gothic vision still looms over popular debate about biotechnology. Simple version – man plays God; creates monster; monster goes on rampage; creator and monster both die (see below under films). But the novel is much more complex and interesting than that. Victor Frankenstein is an idealist, embracing modern science in a quest to overcome death and disease. His creature is intelligent, and engages our sympathy as a newborn reject: Shelley's nightmare is about a motherless child as well as a meditation on life and death. Read online at www.literature.org/authors/shelley-mary/frankenstein (and numerous other sites).

The Island of Dr Moreau H.G. Wells (1896) A bit of *Frankenstein*, some 19th century vivisection, a dose of Darwinian evolution – blend with the darkest imaginings of one of the great masters of the English language and you get that rarest of things, a book which seems to create its own archetypes. It is still hard not to see the crazed Dr Moreau as a genetic engineer before his time: "It is a possible thing to transplant tissue from one part of an animal to another, or from one animal to another, to alter its chemical reactions and methods of growth ... indeed to change it in its most intimate structure." It leaves the reader with an astonishing array of images. You can read Moreau in half an evening. You won't forget it. Read online at www.bartleby.com/1001

Brave New World Aldous Huxley (1932) The opening scenes of Huxley's much-cited novel depict a tour of the hatcheries where future citizens are reared in artificial wombs. Each is pre-destined for their future role as an alpha or an epsilon in a horrifically regimented society. Describing a deeply unattractive culture in realistic detail, while cleverly satirising his own, Huxley created one of the two most widely read 20th century dystopias – only Orwell's *1984* rivals it in influence. The title, taken from Shakespeare's *The Tempest*, has been a gift to headline writers ever since. Read it online at www.huxley.net/bnw

Stand On Zanzibar John Brunner

(1968) British SF writer Brunner showed Atwood how it could be done more than thirty years earlier. A massive, memorable, furiously inventive book, it's impossible to summarise. But one of the many strands Brunner weaves in is the global politics of reproduction. His panoramic novel of a possible future, inspired in part by the vision of the US presented by the 1930s US writer John Dos Passos, mixes real and made-up news reports, letters, diaries, and conventional narrative. The title is derived from the idea that the global population could all crowd onto the island of Zanzibar.

Joshua, Son Of None Nancy

Freedman (1973) Another early cloning novel, an intriguing complement to Levin. This time the tissue fragments come from the dying John F. Kennedy, and the single clone, the biblically named Joshua, is raised in the belief that he is the dead President's illegitimate son. Freedman has Joshua raised in circumstances which repeat episodes from JFK's boyhood – a spinal injury, a swimming mishap. When Joshua discovers the truth he declares himself a new individual, not a replica of his clone father – though he decides that, hey, he'll run for President anyway. To no one's surprise he wins a huge victory. The closing scenes bring a twist which some may guess, but should not

be given away. Suffice to say it adds another layer of ambiguity.

The Boys From Brazil Ira Levin (1976)

Another of those books more often cited than read, usually by journalists trying to evoke the terrors of mass cloning. It has all the right ingredients – evil dictator, creepy scientist, vast conspiracy – as does the film of the same name. Simon Wiesenthal-like Nazi hunter Lieberman (who is played by Laurence Olivier in the movie) tracks down Joseph Mengele, the mad doctor of Auschwitz, to his hideaway in South America. He's been busy, using tissue from the long dead Adolf Hitler to clone babies. The 94 little Hitlers were adopted by childless couples chosen to match the late dictator's parental profile. But when Mengele is eliminated, Lieberman lets the children live. It takes more than genes to make a Hitler, he tells his anxious cohorts; in the end, it was society wot dunnit. A compelling thriller, it has a more subtle view of nature, nurture and history than it is usually given credit for.

Where Late The Sweet Birds Sang

Kate Wilhelm (1976) Two of science fiction's most interesting female authors turned to human cloning in 1976. Pamela Sargent's *Cloned Lives* was a realistically drawn near-future story following the five cloned daughters of brilliant astrophysicist and polymath Paul Swenson. They have to cope with a world of fundamentalist cults, public revulsion, and legislative restrictions on biology labs. Sargent's book would be listed here in its own right but is out of print – though it can be downloaded from www.fictionwise.com/ebooks/eBook1011 and other e-book sites.

Kate Wilhelm's story tells of a future Earth in which humans have lost the power to procreate because of the poisons they have poured into the environment. The only children in this post-*Silent Spring* vision are born through cloning, but they are uneasy with their conventionally born elders. They prefer the company of their fellow clones, who understand each other so much better. The view that this is Wilhelm's best novel is supported by the Hugo award it won from fellow SF authors in 1997. It has just been reissued in the UK (2006) in the SF Masterworks series, and justifies that rubric all through.

The Cloning Of Joanna May Fay Weldon (1989) Joanna May, divorced
from megalomaniac Carl, loses her lover to a contract killer her ex-husband probably paid for. Imagine her surprise when she finds that Carl had also nicked four of her eggs thirty years earlier, under the guise of an abortion, and had her cloned. He really liked her a lot back then, you see. The plot which then ensues is satisfyingly twisty, and conveyed in Weldon's humourously knowing style.

The 60-year-old Joanna eventually meets all the other women: "my sisters, my twins, my clones, my children". As in much of Weldon's output, the humour tempers a bleak vision of relations between men and women and, to a lesser extent, mothers and daughters. As one character, voicing what sounds like the author's view, puts it: life is so fucked up normally, why would genetics make it any worse? The book was entertainingly filmed for British TV in 1992 by Philip Saville.

Mendel's Dwarf Simon Mawer (1997)
Vertically challenged Dr Ben Lambert has achondroplasia, is a brilliant geneticist, and – surprise – is a descendant of Gregor Mendel. With lots of science and a sprinkling of sex, Mawer explores what unites 19th and 20th-century scientists. While Dr Lambert searches for the gene which made him, a parallel story portrays his ancestor, dusting pollen on sweet peas. They are, we gather, pursuing the same project – controlled breeding. But Lambert has more powerful techniques to realize his goals, including screening fertilized embryos so that his lover can bear his child.

Oryx And Crake Margaret Atwood (2003) We're doomed, I tell you
– doomed, all doomed! That's pretty much the message of Canadian Atwood's most recent foray into the fictional future. As well as unfettered genetic engineering of plants, animals and humans, there's global warming, child slavery, Internet porn, and people scratching a living outside, or living high inside secure biotech industry compounds (one is not-so-subtly named Watson-Crick). Oryx And Crake got lots of attention because a noted novelist had picked up some well-worn sci-fi themes and added a literary flourish. But the "pigoons" – intelligent and entertainingly vicious super pigs – stick in your mind. If they turn out to be what is left picking over the rubble, we can't say she didn't warn us.

Sims F. Paul Wilson (2003) How would
we treat enhanced chimps who were intelligent enough to do factory work, wait at table, caddy for golfers, or work as farm hands? An efficiently written thriller, Doherty's book is an interesting exploration of similar territory to Ishiguro's Never Let Me Go, complete with organ farming and enforced prostitution. The plot thickens a tad predictably, and characters ask questions such as: "what makes a guy worthy of your love – his genome or his values?"

Never Let Me Go Kazuo Ishiguro
(2005) Ishiguro's atmospheric narrative is set in an alternative present in which clones are raised as donors whose tissue enables treatment of disease. After four donations, if they are lucky, they "complete", that is, they die. The story is told by a grown-up clone, Kathy, who is still a carer, helping donors to recover after their last operation. As Kathy tries to make sense of her life, and the lives

of those she was raised with in the idyllic but sinister school-cum-community of Hailsham, Ishiguro's characteristically low-key and precise style unveils a horribly memorable picture of a soci-

ety based on exploitation of a whole class of people who everyone else tells themselves are not quite human. A highly readable and chilling feat of imagination.

Further reading

There is quite simply an enormous number of books about genes, genomes and genetics. As far as we know all these are still in print, except where noted, but we suggest you refer to Amazon for latest details of availability. They are listed here in order of first publication.

The Double Helix James Watson (1968) Double-helix discoverer recounts what it was really like in 1953, and upsets all the other people who were there at the time. An unreliable memoir, but hugely readable nonetheless.

The Logic Of Life: A History Of Heredity Francois Jacob (1974) A classic essay on the evolution of ideas about heredity, from the French Nobel laureate and co-discoverer of the mechanism of gene regulation in bacteria. Translated from the French, and beautifully written throughout. It's out of print, but it really shouldn't be.

The Selfish Gene Richard Dawkins (1976) The young Dawkins' classic account of what every gene really wants – controversial and hugely influential, on students, scientists and other science writers.

The Eighth Day Of Creation Horace Judson (1979) Journalist-turned-historian Judson caught up with the pioneers of molecular biology while they were all still alive and wrote this amazing chronicle.

The Extended Phenotype: The Long Reach Of The Gene Richard Dawkins (1982) That man again, with his most technical book. But if you think he overemphasizes the influence of genes,

read chapter 2, which is one of the best accounts of what even he calls the "myth" of genetic determinism.

Invisible Frontiers: The Race To Synthesize A Human Gene Stephen Hall (1987) The story of recombinant human insulin, and the riveting account of high stakes and high drama at the dawn of the new biotechnology.

The Language Of The Genes Steve Jones (1994) The British genetics professor's first popular book, and still a good place to start an exploration of genes and their significance.

Making PCR: A Story Of Biotechnology Paul Rabinow (1996) The author is an anthropologist, so describes his investigation into the origins and development of the all-powerful DNA lab tool that is the polymerase chain reaction in a low-key fashion.

River Out Of Eden Richard Dawkins (1996) One of Dawkins' most accessible books, it presents the history of life as a continuous flow of information.

The DNA Mystique Dorothy Nelkin & Susan Lindee (1996) Wherein a sociologist and a historian analyze DNA in the media, and the culture of gene talk. Two of the most readable academics ever to have commented on this.

The Lives To Come Philip Kitcher (1997) British-born philosopher of science Kitcher spent a year exploring genetics and eugenics and wrote this book outlining how to think about it.

Lifelines Steven Rose (1998) British biologist Rose is a critic of the "genes 'R' us" view of life, and this is his best account of how a genome is only the start of a lifetime of development.

Design For A Life: How Behaviour Develops Patrick Bateson & Paul Martin (1999) An absorbing look at the biology of development by one old, one young scientist.

Genome: The Autobiography Of A Species In 23 Chapters Matt Ridley (1999) Ridley chooses one gene from each chromosome, and tells its story. A clever way into the subject.

Time, Love, Memory: A Great Biologist And His Quest For The Origins Of Behaviour Jonathan Weiner (1999) A brilliant American science writer profiles Seymour Benzer, one of the "lords of the fly", and his work unraveling links between genes and fly behaviour. A great read.

Life Without Genes: The History And Future Of Genomes Adrian Woolfson (2000) A confusing title because it is, in the end, about genes. An imaginative young biologist with a ripe literary style dreams of biologies past and present.

Chance In The House Of Fate: A Natural History Of Heredity Jennifer Ackerman (2001) An American science writer's stylish guide to the unity and diversity of genes across all of life. Full of strange facts and diverting stories.

Dinner At The New Gene Café Bill Lambrecht (2002) Lambrecht, an American journalist, reports from world-wide investigation of the present and future of GM crops and agriculture. A vivid, balanced and informative book

from the midst of a controversy. Good on the differences in public reactions in Britain, Europe and US.

In Mendel's Footnotes Colin Tudge (2002) A leading British science writer writes a history of genetics as a series of footnotes to Mendel (following the dictum that philosophy is said to be footnotes to Plato). This one would make a real historian blanch, but the result is fascinating.

In The Beginning Was The Worm Andrew Brown (2003) A whole book about a worm? Well, Darwin wrote one. British journalist Brown focuses on the tiny nematode worm on which genome pioneers such as John Sulston built their careers.

The Common Thread: A Story Of Science, Politics, Ethics And The Human Genome John Sulston & Georgina Ferry (2003) Mild-mannered Englishman John Sulston devotes life to worm, wins Nobel prize, and tells us all about it with the aid of skilful science writer Ferry.

Nature Via Nurture Matt Ridley (2003) Ridley looks at the subtleties of the interaction between genes, experience and environment, and how gene expression is constantly modified by these influences. Good on the latest research, though some of his claims are overstated.

Pointing From The Grave: A True Story Of Murder And DNA Samantha Weinberg (2003) A science writer's true-crime style reconstruction of a single case, which offers a painless way to learn about DNA fingerprinting and how it is used in court.

Genes And Behaviour: Nature-Nurture Interplay Explained Michael Rutter (2006) A British psychologist's judicious review of all the latest information on how genes shape what we do.

Genes on screen

Here are ten movies that yield images or scenes which may find a lasting place in popular consciousness of genetics and cloning. For more details of 20th century films of this kind *Screening DNA: Exploring the Cinema-Genetics Interface*, by Stephen Nottingham. This book-length study never made it into print but is published on the web at ourworld.compuserve. com/homepages/Stephen_Nottingham/DNA1

Frankenstein James Whale (1931)
In the introduction to a revised edition of her novel in 1831 Mary Shelley referred affectionately to her "hideous progeny". They have indeed multiplied, and nowhere more so than in the cinema. The *Illustrated Frankenstein Movie Guide* in 1994 listed over four hundred productions loosely based on the story. The English director James Whale made his classic black-and-white horror vision of *Frankenstein* for Universal in 1931. This was the first screen appearance of Boris Karloff as the iconic speechless creature: stiffly dignified, curious, and full of pathos. It has the most manic of scientists, and a creation scene even more powerful than Mary Shelley's original. All together now: "It's alive!!!"

Blade Runner Ridley Scott (1982)
The classic vision of a grungy future, *Blade Runner* depicts world-weary Harrison Ford as Deckard, hunting down escaped off-world "replicants" – industrially produced android slaves with perfect physiques but brief life spans. The doomed replicants are trying to reach their creator, the corporate genius Tyrell, to demand "more life". This is the best-looking, the most complex and the most intriguing of the films listed here. The director's cut, shorn of the unnecessary narrative voice-over, is correctly hailed as superior to the original cinema release. An enduring classic.

The Fly David Cronenberg (1986)
Cronenberg's take on this 1950s scenario – man with matter transmitter gets mixed up with insect – is that rare thing, a superior remake. As noted earlier, there are gestures toward modern science in the idea that the machine has somehow integrated the genes of the two creatures. But the magic of the film is in the compelling central performance from Jeff Goldblum as he comes to comprehend his awful fate. Philosopher Thomas Nagel famously asked, "what is it like to be a bat"? Cronenberg and Goldblum show something of what it might be like to be a fly. Not at all nice.

Jurassic Park Steven Spielberg (1993)
Michael Crichton's novel read like a film treatment, and Spielberg took it to the screen with awe-inspiring special effects and a complete scrambling of the more or less coherent take on chaos and systems theory of the book. Never mind – it's the dinosaurs that are the stars, and they run amok in convincingly species-specific ways. You'll believe that an extinct creature can be recreated from DNA, and there's even a cute educational cartoon along the way to explain how. It has probably been seen by more people than any other explanation of what the master molecule can do and how it works.

Gattaca Andrew Nicol (1997)
A touch of *Brave New World* about the storyline, but this future certainly has an original look – albeit one somehow rooted in the 1950s. The moderately compelling plot centres on the struggle of one of the genetically "in-valid" to escape the underclass and become an

astronaut. He fixes up an identity swap with a genetically perfect specimen who has been crippled in an accident. The feelgood ending carries the message that in a society built around the idea that DNA is destiny "there's no gene for the human spirit". The disabled owner of the new astronaut's original identity, however, remains firmly earthbound – leaving a rather unpleasant aftertaste.

The Sixth Day Roger Spottiswoode (2000) You just knew Arnold Schwarzenegger would get in here somewhere. He comes home one day to find he has been cloned, a teensy slip-up by a bunch of corporate conspirators who thought he died in a helicopter crash. The interest of this one, as in several of these selections, is in the details of the envisioned future. Biotech has abolished world hunger, and is used to clone organs for transplant and replace dead pets. But it is kept under a tight rein, which some want to throw off to achieve a kind of immortality.

Mission To Mars Brian da Palma (2000) Worth noting for the idea, rather than the images. This is a routine sub-*2001* space movie, in which commander Luke Graham leads the first manned mission to Mars. His crew happen upon an ages-old domelike alien artefact. It activates, and kills everyone but him. When the rescue mission arrives, the recovering Graham has fathomed the history of the dome – property of an ancient Martian civilisation which first brought life to Earth. How did they do it? Why, they sent DNA, of course, which did the rest all by itself.

Code 46 Michael Winterbottom (2003) Another grungy future, this time abounding in so many clones that the government has regulated reproduction – no procreation with anyone who shares 25 percent or more of your genes. So you might be able to guess what will be uncovered when investigator Geld finds himself strangely attracted to a woman using fake ID papers that he has been hired to pursue. The film has a dark mix of dystopian elements – authoritarianism, globalisation, and environmental deterioriation. But the doomed love story is the heart of the picture. Tim Robbins and Samantha Morton did not move all the critics, but it is well worth seeking out.

Godsend Nick Hamm (2004) "The Godsend Institute is a fertility clinic and practice, specializing in the replication of cells for the purpose of creating life from life ... we have perfected a procedure by which a single cell could be used to create a genetically identical foetus." This is the introduction to the Godsend Institute on the rather clever fake website set up to publicize the movie. Despite the presence of Robert De Niro as the doctor who presides over human cloning, the website is the only clever thing about the movie. It has a weak and clichéd story – clone baby turns into devil child. But the web stunt is still an intriguing example of how fictional products with a realistic sheen can play with our expectations of future science and technology.

The Island Michael Bay (2005) A would-be blockbuster with an interesting premise: how would you keep alive (and fit and healthy) a group of clones who were destined only to be organ donors? The film's answer is a comfy but confined futuristic community, in which the doomed residents are told they can only leave to visit 'The Island', which they are told is the only other habitable spot on a ravaged planet. Lincoln Six-Echo (Ewan McGregor having some well-paid fun) discovers the truth, and escapes to seek a life where he will not be "harvested" when his parts are needed. Cue a switch to a spectacular but vapid action movie, a sort of more energetic *Logan's Run*.

Genes online

There are many websites featuring topics in this book worth visiting. We have mentioned some in sections relating to specific topics. Here is a selection of other, more general sites, which are good starting points for further inquiry.

www.actionbioscience.org Assembled by the American Institute of Biological Sciences, this site has resources on many topics, including genomics, evolution, biotechnology and biodiversity. Most are well-chosen articles reproduced from good sources.

www.bionews.org.uk Archive site for invaluable weekly genetics, reproduction and embryo research newsletter, produced by UK charity Progress Educational Trust.

www.yourgenome.org Education site from the Wellcome Trust Sanger Institute in Cambridge (not to be confused with the separate Wellcome Trust genome site – see below). It's usefully divided into "beginner", "intermediate" and "advanced" sections. But they haven't got round to being advanced yet, and the rest is showing its age…

www.dnapolicy.org Useful US site maintained by the Genetics and Public Policy Center of Johns Hopkins University.

www.cancer.gov/cancertopics/understandingcancer Another US site, this time from the National Institutes of Health. Has guides to cancer genomics, genetic variation and, especially useful, genetic testing.

talk.dnadirect.com A wide-ranging blog supported by a DNA testing company. Well worth dipping into for recent news and commentary.

www.geneticseducation.nhs.uk Intended for staff working in the UK National Health Service, but there's no reason why the rest of us can't read it. Good on individual genetic diseases.

www.phgu.org.uk Site of the Public Health Genetics Unit, an academic and policy outfit in Cambridge, UK. Very good news archive and collection of articles and papers.

www.cdc.gov/genomics Public Health Genomics in the USA.

genome.wellcome.ac.uk Nice, well-resourced education site, run by the UK Wellcome Trust.

www.genome.gov With a URL like that, they had to be early in the game. This is the US National Institutes of Health Human Genome Research Institute. Mainly aimed at researchers, but has good health, education and ethics sections for the rest of us.

www.ornl.gov/sci/techresources/Human_Genome US government Genetics virtual library. All about the genome: boy, is this comprehensive.

Glossary

Adult stem cell A type of cell found in adult tissues that is capable of dividing and multiplying to produce a range of different specialized cells. For example, bone marrow stem cells produce a wide range of different blood cells.

Allele Alternate forms of a gene at a particular chromosome location – the gene governing ABO blood type, for example – or indeed any alternative DNA sequences at a particular chromosome location.

Amino acid The chemical building blocks that make up proteins. There are 20 different common amino acids.

Amniocentesis A test carried out at or after the 15th week of pregnancy, in which a small amount of the fluid that surrounds the foetus is removed for biochemical or genetic testing.

Aneuploidy A condition in which there is an abnormal number of chromosomes, either fewer or more than is usual (46 in a human body cell).

Anticipation The tendency of some genetic conditions to increase in severity – eg myotonic dystrophy – or appear earlier – eg Huntington's disease – in successive generations of the same family.

Association The observed connection between an inherited trait and a particular gene or DNA marker. Association studies are used to pinpoint genes involved in complex disorders which are influenced by several genetic and non-genetic factors.

Atom The basic component of all matter. The atom is the smallest particle of a chemical element that has all of the properties of that element.

Autosomal dominant inheritance A pattern of inheritance in which a gene change has an effect even if only one copy is inherited, eg the mutated gene that causes Huntington's disease. Also known simply as dominant inheritance.

Autosomal recessive inheritance A pattern of inheritance in which a gene change has an effect only if two copies are inherited – the mutated gene that causes cystic fibrosis, for instance. Also known simply as recessive inheritance.

Autosome Any human chromosome apart from the sex chromosomes, X and Y. There are 22 pairs of autosomes, numbered 1–22.

Base A building block of DNA; a nitrogen-based molecule which, when paired with another base, forms one "rung" of the DNA double helix. In DNA, the base adenine (A) always pairs with thymine (T), and cytosine (C) always pairs with guanine (G).

Base-pair (bp) Two bases paired together by weak chemical bonds. Lengths of DNA are referred to in terms of the number of base-pairs they contain, eg 250bp.

Blastocyst A mammalian embryo in the first stage of development, when the fertilised egg has grown into a hollow ball made up of a few hundred cells. In humans, the blastocyst stage lasts from the fourth to eleventh day following fertilisation.

Carrier An individual who has a disease-causing gene mutation on one chromosome of a pair, and a normal version of the gene on the other. Usually refers to unaffected carriers of recessive conditions.

Cell Cells are the building blocks of all living things. A human body is made up of around 100 million million cells.

Chemical Anything made of matter is a chemical, although the word is generally used to describe pure substances, rather than mixtures.

Chemical element A pure substance composed of identical atoms, for example iron or oxygen. Elements cannot be reduced to simpler substances by normal chemical means.

Chimera An individual with a mixture of body cells that have two different sets of genetic information, arising when two early embryos fuse together in the womb.

Chorionic villus sampling (CVS) A test carried out at or after the 10th week of pregnancy, in which a few cells of the placenta are removed for biochemical or genetic testing.

Chromosome Chromosomes are tightly packaged bundles of DNA, the chemical that encodes genetic information. Nearly all human body cells have a set of 46 chromosomes, while egg and sperm cells have 23.

Clone 1. An exact copy of a piece of DNA (which may or may not be a gene), isolated and propagated in a bacterial or yeast host for further study. 2. An exact genetic replica of another living organism or cell.

Congenital Present from birth. A congenital condition may or may not have a genetic basis.

Cytogenetics The study of chromosomes, usually to detect abnormalities associated with a medical condition.

Cytoplasm The jelly-like material present inside a eukaryotic cell (eg plant and animal cells) which surrounds the nucleus.

Deletion A missing portion of genetic material, from within either a gene, section of DNA, or piece of chromosome.

DNA Deoxyribonucleic acid (DNA) is the chemical that encodes genetic information. It contains four different chemicals, or bases, known as A (adenine), C (cytosine), G (guanine) and T (thymine).

DNA marker A piece of DNA that varies between individuals. DNA markers are used to track the inheritance of disease genes, for both clinical and research reasons.

DNA probe A piece of DNA, labelled either chemically or with radioactivity, used in the laboratory to detect a specific gene, gene mutation or DNA marker.

DNA profiling/fingerprinting A laboratory technique used to match samples of DNA, in forensics for example, or to establish the relationship between two people, as in paternity testing.

Duplication A duplicated portion of genetic material, either a gene, section of DNA, or piece of chromosome.

Embryo A stage of development (before the foetus stage) during which tissues and organs are formed. In humans, this stage lasts for eight weeks after the fertilised egg first starts to divide.

Embryonic stem (ES) cell A type of unspecialized cell found in early embryos, which is capable of developing into all (multipotent) or a wide range (pluripotent) of different body tissues.

Enzyme Proteins that control the speed of chemical reactions that take place in living things.

Eukaryote Organisms made up of complex cells, in which the vast majority of the genetic material is contained within a membrane-bound nucleus. All animals, plants and fungi are eukaryotes

Epigenetic Refers to mechanisms that control gene activity, which do not change the DNA sequence itself, for example, DNA methylation.

Eugenic Refers to methods aimed at improving the health of human populations through the application of genetics.

Exon Exons are the parts of a gene that are included in the final messenger RNA copy of the DNA sequence, which is then used to make a particular protein. In a gene, exons are interspersed with introns, which are "edited out" during this process.

FISH Fluorescent in situ hybridization (FISH) is a laboratory technique that uses fluorescently-labelled pieces of DNA to detect specific genes, chromosome segments or chromosomes, in cells examined under a special microscope.

Foetus A stage of development (after the embryo stage) when all organs have been formed. In humans, this stage lasts from nine weeks after fertilisation until birth.

Gene The basic biological unit of inheritance. Genes are made out of DNA, and most are coded instructions for making proteins. Others make molecules that control gene activity.

Gene therapy An experimental medical technique that aims to treat an illness by replacing a faulty or missing gene with a working copy, or by switching off a harmful gene.

Genetic condition A condition or illness caused by changes in a gene or genes, affecting the way the body looks, develops or works.

Genetic counselling Information and advice given to patients affected by, or at risk of a genetic condition. An explanation of risks and options may include the findings of specific genetic tests.

Genetic modification/engineering The permanent alteration of an organism's genetic material, using laboratory-based techniques rather than conventional breeding methods.

Genetic susceptibility/predisposition Increased probability (compared to the general population) of developing a disease, due to the presence of one or more gene mutations.

Genetic test A test that gives genetic information, for example about paternity, disease status or disease susceptibility.

Genome The total genetic information of a living thing, a complete copy of which is found in most body cells. The human genome is made up of around 2.9 billion chemical letters (base-pairs) of DNA.

Genotype The specific set of gene variants (alleles) inherited at a particular chromosome location.

Germ cells The reproductive cells (egg and sperm), and their precursors.

Haplotype A set of several different DNA or gene variants present together in a particular section of a chromosome.

HapMap An international project looking at genetic variation in people from around the world, to speed the search for genes linked to common diseases. The scientists are focussing on genetic markers (called single nucleotide polymorphisms, or SNPs) that have been inherited together in a "haplotype block" over many generations.

Heritability The extent to which a particular trait is inherited. A trait that is entirely genetic (eg eye colour), has a heritability of one, whereas something entirely environmental (eg a tattoo) has a heritability of zero.

Heterozygote An individual who inherits two different alleles (one from each parent) at a particular chromosome location. When referring to genetic disorders, a heterozygote is usually someone who has one disease and one non-disease allele.

Homozygote An individual who has inherited two identical alleles at a particular chromosome location. When referring to genetic disorders, a homozygote is usually someone who has two disease alleles.

Hormone A chemical messenger produced in the body – for example testosterone. Hormones can travel through the blood and have an effect elsewhere in the body.

Human Genome Project (HGP) An international, publicly funded effort to read and decode the entire genetic information of a human being, the results of which were published in April 2003.

Imprinting The process by which certain mammalian genes are "switched off", according to whether they were inherited from the father or mother.

Inbreeding Mating between closely related individuals in a population, eventually resulting in a decrease in the total amount of genetic variation present.

Intron Non-coding parts of a gene that are "edited out", and so do not contribute to making the final protein product of the gene. Introns are scattered between the exons of a gene.

Insertion A portion of genetic material inserted into either a gene, a section of DNA, or a piece of chromosome.

Inversion A type of chromosome rearrangement, in which a section of chromosome is removed, turned upside down and reinserted back into its place in the chromosome.

IVF IVF (in vitro fertilisation) is a treatment for infertility, in which eggs are removed from a woman's body, fertilised with sperm in a laboratory, then returned to the womb shortly afterwards to continue developing.

Karyotype A photograph of the chromosomes of a single cell, as seen down a microscope, cut out and arranged in pairs, based on their size and banding pattern after chemical staining.

Linkage analysis A genetic technique used to track the inheritance of a genetic condition through a family, using genetic markers closely linked to the suspected disease gene.

Meiosis The form of cell division by which egg and sperm cells are formed. Unlike normal cell division (mitosis), the number of chromosomes in the resulting cells is half that in the original cells.

Mendelian inheritance A pattern of inheritance displayed by a trait under the control of one gene, which fits one of those first described by Gegor Mendel.

Messenger RNA Messenger RNA (mRNA) is an intermediate stage between a gene and the protein it codes for. The cell uses an mRNA template when making a protein, rather than reading the DNA code directly.

Microarray An array of many different pieces of DNA, arranged on a small piece of glass, silicon or other solid material. Microarrays, also known as "DNA chips", can be used to study thousands of different genes or genetic variations at the same time.

Microsatellite Repeated sections of DNA, two to five base-pairs in length, found scattered throughout the human genome. Because they are highly variable, microsatellites are often used in genetic studies, and in forensic and paternity tests.

Mitochondria Energy-generating structures found in eukaryotic cells (eg plants and animals) which contain a few genes.

Mitochondrial inheritance An unusual pattern of inheritance caused by mutations in genes found in the mitochondria. Mitochondrial genes are always maternally inherited, since sperm contribute no mitochondria to the fertilized egg.

Mitosis The ordinary form of cell division, in which a cell divides to produce two new cells that both have the same number of chromosomes as the original cell.

Molecule The smallest particle of a substance that has all of the physical and chemical properties of that substance. Molecules are made up of two or more atoms, for example, a molecule of water is made up of two atoms of hydrogen and one of oxygen.

Monosomy The presence of only one chromosome from a pair in an individual. With the exception of sex chromosomes (eg Turner syndrome), monosomies usually have severe effects, usually resulting in early pregnancy loss.

Mosaicism The occurrence of two or more cell populations within a single tissue or individual, which have different genetic constitutions.

Mutation A change in the genetic information of an individual, which may have a harmful, neutral or beneficial effect.

Nucleotide Nucleotides are the building blocks that make up DNA and RNA molecules. A single nucleotide consists of a nitrogenous base (adenosine [A], cytosine [C], guanine [G] or thymine [T] in DNA), a phosphate group, and a sugar molecule.

Nucleus A structure found in the centre of eukaryotic cells (eg plant and animal cells), which contains most of its genetic material.

Parthenogenesis The creation of artificial "embryos" (parthenotes) without the need for fertilisation of an egg by sperm.

Pedigree A diagram of the genetic relationships and medical history of a family, which uses standard symbols to represent affected and unaffected males and females.

Penetrance The extent to which a mutation causing a particular disorder causes clinical symptoms of that disorder. Usually refers to autosomal dominant conditions.

Pharmacogenetics An area of research that aims to develop ways of matching medicines to a person's genetic make-up, to avoid adverse reactions and non-response to a particular drug.

Phenotype The physical characteristics caused by a particular gene variant, or combinations of gene variants (alleles).

Polymerase chain reaction (PCR) A laboratory technique used to make millions of copies of a known piece of DNA, used in many medical and forensic tests.

Polymorphic Used to describe a natural variation in a gene, piece of DNA, protein, or chromosome that has no effect on the individual who inherits it.

Pre-implantation genetic diagnosis PGD is a genetic test that can be carried out on embryos created using *in vitro* fertilisation (IVF), to ensure that only embryos unaffected by a particular genetic condition are returned to the woman's womb.

Prenatal diagnosis Biochemical, genetic or ultrasound test performed during pregnancy, to determine if a foetus is affected by a particular disorder.

Prokaryote A simple organism, usually single-celled, which lacks a cell nucleus and other membrane-bound compartments. All bacteria are prokaryotes.

Protein A large biological molecule made up of a string of sub-units called amino acids. There are thousands of different proteins in the human body, each with a particular job. Haemoglobin, for example, carries oxygen around the blood.

Recombination The exchange of a segment of DNA between two chromosomes of the same pair during meiosis, leading to a novel combination of genetic material in the egg or sperm.

Recombinant DNA technology The creation of new DNA molecules using fragments of DNA from different sources, which can then be used to make genetically modified organisms.

Ribonucleic acid (RNA) A close chemical relative of DNA, which plays important roles in controlling gene activity and decoding gene instructions. It contains four different chemicals, or bases, known as A (adenine), C (cytosine), G (guanine) and uracil (U).

RNA interference (RNAi) A naturally-occurring cell process that is being exploited as a way of selectively shutting down gene activity. It involves injecting cells with short, specific pieces of RNA, which then trigger the breakdown of a particular messenger RNA molecule.

Sequence The order of base-pairs along a stretch of DNA, or the determining of this order.

Sex chromosome Refers to the two chromosomes (X and Y) that determine the sex of an individual.

Single gene disorder A disorder caused by a mutation in a single gene. Cystic fibrosis, for example, is caused by mutations in the CFTR gene.

Splicing The removal of introns from the messenger RNA copy of a gene, so that just the gene exons are present in the final version.

SNP An SNP (single nucleotide polymorphism) is a single chemical unit (base-pair) variation in a stretch of DNA.

Somatic Refers to all body cells apart from the egg and sperm cells, and their precursors.

Somatic cell nuclear transfer (SCNT) A technique in which the genetic material from an egg cell is replaced with that of an adult or embryo body cell of the same animal species. Also known as cell-nucleus replacement (CNR), or more popularly as "cloning".

Telomere The segment at both ends of each chromosome, consisting of a series of repeated DNA sequences. Some of the telomere is lost each time a cell divides, and eventually, when it is gone, the cell dies.

Therapeutic cloning A popular name for the proposed use of embryo stem cells, derived using somatic cell nuclear transfer technology, to develop genetically matched cell therapies for a range of diseases.

Transgenic A genetically altered organism that contains genes from another species.

Trisomy The presence of an extra chromosome, resulting in a total of three chromosomes of that particular type instead of a pair, eg trisomy 21, which causes Down syndrome.

UK BioBank A government-funded project to collect DNA samples and medical records from 500,000 British volunteers aged 45–60, to study the effects of genetic and environmental factors on health.

Vector A delivery system used in the laboratory to carry foreign DNA into a cell. Modified viruses are commonly used as vectors in gene therapy experiments.

Virus A tiny infectious particle that can invade a cell.

X-chromosome One of the two types of sex chromosome, X and Y. In humans, females have two X-chromosomes, while males have one X and one Y.

X-inactivation The process in female mammals by which one X-chromosome is randomly inactivated in the cells of the early embryo.

X-linked inheritance, dominant A pattern of inheritance caused by a gene mutation located on the X-chromosome, which has an effect even if only one copy is inherited, so it affects girls (who have two X-chromosomes) as well as boys (who have one X-chromosome).

X-linked inheritance, recessive A pattern of inheritance displayed by a gene mutation located on the X-chromosome, which has an effect if one copy is inherited by a boy (who have only one X chromosome). Girls are usually unaffected, or only carriers of recessive X-linked conditions.

Xenotransplantation The transplanting of animal organs into human patients.

Y-chromosome One of the two types of sex chromosome, X and Y. In humans, females have two X-chromosomes, while males have one X and one Y.

Zygote The single cell formed immediately after fertilisation, once the genetic information of the egg and sperm have fused.

Index

Picture credits

Corbis: 10, 43, 72, 91, 148, 211, 237, 238, 263, 266; Jeff Miller/University of Wisconsin-Madison: 235; Jupiter: 9; National Cancer Institute: 51, 53, 103, 150; National Human Genome Research Institute: 34; National Library of Medicine: 21, 99, 182, 289; Oak Ridge National Laboratory: 73; Oregon State University: 60; Peter Gruber Foundation: 70; Rocky Mountain Laboratories: 183; Smithsonian Institute: 116, 117; University of Melbourne: 27; University of Texas: 56

Colour section Dr Bruce Chassey Laboratory, National Institute of Dental Research: 3; Corbis: 1, 2, 4; Dr Jason Kang, National Cancer Institute: 3; National Human Genome Research Institute: 2; US Department of Energy Human Genome Program: 1

D: Rough Guide
DIRECTIONS for
short breaks

Available from all good bookstores

For more information go to www.roughguides.com

ROUGH GUIDES

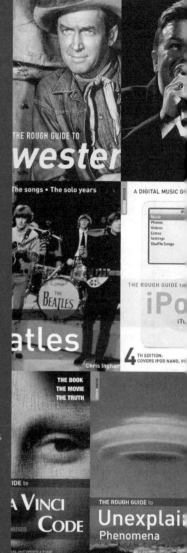